STUDENT'S SOLUTIONS MANUAL

SUSAN JANE COLLEY

VECTOR CALCULUS

FOURTH EDITION

Susan Jane Colley

Oberlin College

PEARSON

Boston Columbus Indianapolis New York San Francisco Upper Saddle River
Amsterdam Cape Town Dubai London Madrid Milan Munich Paris Montreal Toronto
Delhi Mexico City Sao Paulo Sydney Hong Kong Seoul Singapore Taipei Tokyo

Reproduced by Pearson from electronic files supplied by the author.

Publishing as Pearson, 75 Arlington Street, Boston, MA 02116.

ISBN-13: 978-0-321-78067-6
ISBN-10: 0-321-78067-1

15 2021

www.pearsonhighered.com

PEARSON

Contents

Chapter 1

Vectors

1.1 Vectors in Two and Three Dimensions

3. **(a)** $(3,1)+(-1,7)=(3+[-1],1+7)=(2,8)$.
(b) $-2(8,12)=(-2\cdot 8,-2\cdot 12)=(-16,-24)$.
(c) $(8,9)+3(-1,2)=(8+3(-1),9+3(2))=(5,15)$.
(d) $(1,1)+5(2,6)-3(10,2)=(1+5\cdot 2-3\cdot 10,1+5\cdot 6-3\cdot 2)=(-19,25)$.
(e) $(8,10)+3((8,-2)-2(4,5))=(8+3(8-2\cdot 4),10+3(-2-2\cdot 5))=(8,-26)$.

5. We start with the two vectors **a** and **b**. We can complete the parallelogram as in the figure on the left. The vector from the origin to this new vertex is the vector $\mathbf{a}+\mathbf{b}$. In the figure on the right we have translated vector **b** so that its tail is the head of vector **a**. The sum $\mathbf{a}+\mathbf{b}$ is the directed third side of this triangle.

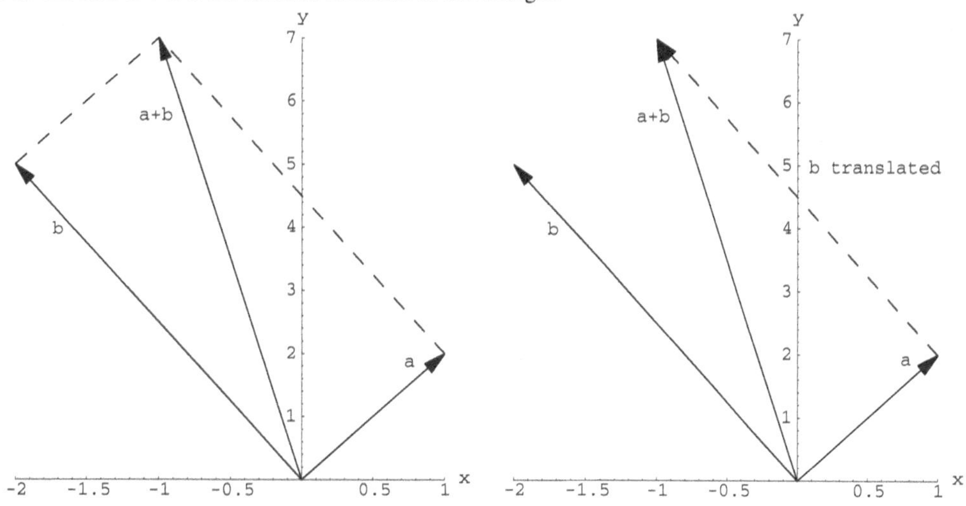

9. *Since the sum on the left must equal the vector on the right componentwise*:
$-12+x=2$, $9+7=y$, and $z+-3=5$. Therefore, $x=14$, $y=16$, and $z=8$.

13. The natural extension to higher dimensions is that we still add componentwise and that multiplying a scalar by a vector means that we multiply each component of the vector by the scalar. In symbols this means that:
$\mathbf{a}+\mathbf{b}=(a_1,a_2,\ldots,a_n)+(b_1,b_2,\ldots,b_n)=(a_1+b_1,a_2+b_2,\ldots,a_n+b_n)$ and $k\mathbf{a}=(ka_1,ka_2,\ldots,ka_n)$.
In our particular examples, $(1,2,3,4)+(5,-1,2,0)=(6,1,5,4)$, and $2(7,6,-3,1)=(14,12,-6,2)$.

17. If **a** is your displacement vector from the Empire State Building and **b** your friend's, then the displacement vector from you to your friend is $\mathbf{b}-\mathbf{a}$.

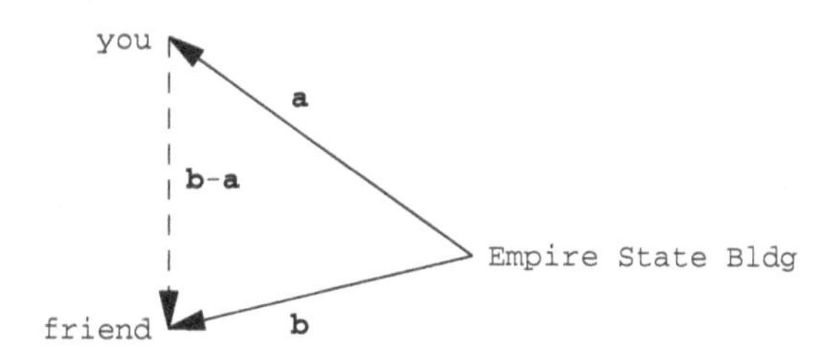

19. *We provide the proofs for* $\mathbf{R}^3$:

$$\begin{aligned}(1)\ (k+l)\mathbf{a} &= (k+l)(a_1,a_2,a_3) = ((k+l)a_1,(k+l)a_2,(k+l)a_3)\\ &= (ka_1+la_1,ka_2+la_2,ka_3+la_3) = k\mathbf{a}+l\mathbf{a}.\\ (2)\ k(\mathbf{a}+\mathbf{b}) &= k((a_1,a_2,a_3)+(b_1,b_2,b_3)) = k(a_1+b_1,a_2+b_2,a_3+b_3)\\ &= (k(a_1+b_1),k(a_2+b_2),k(a_3+b_3)) = (ka_1+kb_1,ka_2+kb_2,ka_3+kb_3)\\ &= (ka_1,ka_2,ka_3)+(kb_1,kb_2,kb_3) = k\mathbf{a}+k\mathbf{b}.\\ (3)\ k(l\mathbf{a}) &= k(l(a_1,a_2,a_3)) = k(la_1,la_2,la_3)\\ &= (kla_1,kla_2,kla_3) = (lka_1,lka_2,lka_3)\\ &= l(ka_1,ka_2,ka_3) = l(k\mathbf{a}).\end{aligned}$$

21. **(a)** The head of the vector $s\mathbf{a}$ is on the x-axis between 0 and 2. Similarly the head of the vector $t\mathbf{b}$ lies somewhere on the vector $\mathbf{b}$. Using the head-to-tail method, $s\mathbf{a}+t\mathbf{b}$ is the result of translating the vector $t\mathbf{b}$, in this case, to the right by $2s$ (represented in the figure by $t\mathbf{b}$*). The result is clearly inside the parallelogram determined by **a** and **b** (and is only on the boundary of the parallelogram if either t or s is 0 or 1.

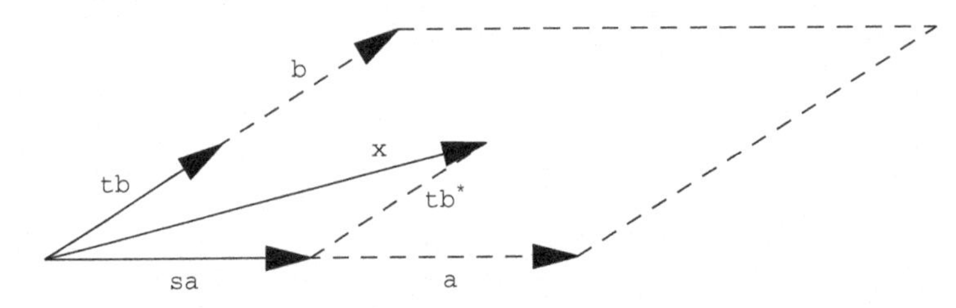

(b) Again the vectors **a** and **b** will determine a parallelogram (with vertices at the origin, and at the heads of **a**, **b**, and $\mathbf{a}+\mathbf{b}$. The vectors $s\mathbf{a}+t\mathbf{b}$ will be the position vectors for all points in that parallelogram determined by (2, 2, 1) and (0, 3, 2).

25. **(a)** Adding we get: $\mathbf{F}_1+\mathbf{F}_2 = (2,7,-1)+(3,-2,5) = (5,5,4)$.

(b) You need a force of the same magnitude in the opposite direction, so $\mathbf{F}_3 = -(5,5,4) = (-5,-5,-4)$.

1.2 More about Vectors

3. $(3,\pi,-7) = 3(1,0,0)+\pi(0,1,0)-7(0,0,1) = 3\mathbf{i}+\pi\mathbf{j}-7\mathbf{k}$.

7. $9\mathbf{i}-2\mathbf{j}+\sqrt{2}\mathbf{k} = 9(1,0,0)-2(0,1,0)+\sqrt{2}(0,0,1) = (9,-2,\sqrt{2})$.

11. **(a)** $(3,1) = c_1(1,1)+c_2(1,-1) = (c_1+c_2,c_1-c_2)$, so $\begin{cases} c_1+c_2=3, \text{ and}\\ c_1-c_2=1.\end{cases}$

Solving simultaneously (for instance by adding the two equations), we find that $2c_1 = 4$, so $c_1 = 2$ and $c_2 = 1$. So $\mathbf{b} = 2\mathbf{a}_1+\mathbf{a}_2$.

(b) Here $c_1+c_2 = 3$ and $c_1-c_2 = -5$, so $c_1 = -1$ and $c_2 = 4$. So $\mathbf{b} = -\mathbf{a}_1+4\mathbf{a}_2$.

(c) More generally, $(b_1,b_2) = (c_1+c_2,c_1-c_2)$, so $\begin{cases} c_1+c_2=b_1, \text{ and}\\ c_1-c_2=b_2.\end{cases}$

Again solving simultaneously, $c_1 = \dfrac{b_1+b_2}{2}$ and $c_2 = \dfrac{b_1-b_2}{2}$. So

$$\mathbf{b} = \left(\frac{b_1+b_2}{2}\right)\mathbf{a}_1+\left(\frac{b_1-b_2}{2}\right)\mathbf{a}_2.$$

17. $\mathbf{r}(t) = (1,4,5)+t(2-1,4-4,-1-5)$ so $\begin{cases} x=1+t\\ y=4\\ z=5-6t.\end{cases}$

21. **(a)** $\mathbf{r}(t) = (-1,7,3)+t(2,-1,5)$ so $\begin{cases} x=-1+2t\\ y=7-t\\ z=3+5t.\end{cases}$

(b) $\mathbf{r}(t) = (5, -3, 4) + t(0-5, 1+3, 9-4)$ so $\begin{cases} x = 5 - 5t \\ y = -3 + 4t \\ z = 4 + 5t. \end{cases}$

(c) Of course, there are infinitely many solutions. For our variation on the answer to (a) we note that a line parallel to the vector $2\mathbf{i} - \mathbf{j} + 5\mathbf{k}$ is also parallel to the vector $-(2\mathbf{i} - \mathbf{j} + 5\mathbf{k})$ so another set of equations for part (a) is:

$$\begin{cases} x = -1 - 2t \\ y = 7 + t \\ z = 3 - 5t. \end{cases}$$

For our variation on the answer to (b) we note that the line passes through both points so we can set up the equation with respect to the other point:

$$\begin{cases} x = -5t \\ y = 1 + 4t \\ z = 9 + 5t. \end{cases}$$

(d) The symmetric forms are:

$$\frac{x+1}{2} = 7 - y = \frac{z-3}{5} \qquad \text{(for (a))}$$

$$\frac{5-x}{5} = \frac{y+3}{4} = \frac{z-4}{5} \qquad \text{(for (b))}$$

$$\frac{x+1}{-2} = y - 7 = \frac{z-3}{-5} \qquad \text{(for the variation of (a))}$$

$$\frac{x}{-5} = \frac{y-1}{4} = \frac{z-9}{5} \qquad \text{(for the variation of (b))}$$

25. Let $t = (x+5)/3$. Then $x = 3t - 5$. In view of the symmetric form, we also have that $t = (y-1)/7$ and $t = (z+10)/(-2)$. Hence a set of parametric equations is $x = 3t - 5$, $y = 7t + 1$, and $z = -2t - 10$.

29. Here again the vector forms of the two lines can be written so that we see their headings are the same:

$$\mathbf{r_1}(t) = (2, -7, 1) + t(3, 1, 5)$$

$$\mathbf{r_2}(t) = (-1, -8, -3) + 2t(3, 1, 5).$$

The point $(2, -7, 1)$ is on line one, so we will check to see if it is also on line two. As in Exercise 28 we check the equation for the x component and see that $-1 + 6t = 2 \Rightarrow t = 1/2$. Checking we see that $\mathbf{r_2}(1/2) = (-1, -8, -3) + (1/2)(2)(3, 1, 5) = (2, -7, 2) \neq (2, -7, 1)$ so the equations do not represent the same lines.

33. We can substitute the parametric forms of x, y, and z into the equation for the plane and solve for t. So $(3t - 5) + 3(2 - t) - (6t) = 19$ which gives us $t = -3$. Substituting back in the parametric equations, we find that the point of intersection is $(-14, 5, -18)$.

37. For points on the line we see that $x - 3y + z = (5 - t) - 3(2t - 3) + (7t + 1) = 15$, so the line does not intersect the plane.

41. We just plug the parametric expressions for x, y, z into the equation for the surface:

$$\frac{(at+a)^2}{a^2} + \frac{b^2}{b^2} - \frac{(ct+c)^2}{c^2} = \frac{c^2(t+1)^2}{a^2} + 1 - \frac{c^2(t+1)^2}{c^2} = 1$$

for all values of $t \in \mathbf{R}$. Hence all points on the line satisfy the equation for the surface.

45. (a) As in Example 2, this is the equation of a circle of radius 2 centered at the origin. The difference is that you are traveling around it three times as fast. This means that if t varied between 0 and 2π that the circle would be traced three times.

(b) This is just like part (a) except the radius of the circle is 5.

(c) This is just like part (b) except the x and y coordinates have been switched. This is the same as reflecting the circle about the line $y = x$ and so this is also a circle of radius 5. If you care, the circle in (b) was drawn starting at the point (5, 0) counterclockwise while this circle is drawn starting at (0, 5) clockwise.

(d) This is an ellipse with major axis along the x-axis intersecting it at $(\pm 5, 0)$ and minor axis along the y-axis intersecting it at $(0, \pm 3)$: $\frac{x^2}{25} + \frac{y^2}{9} = 1$.

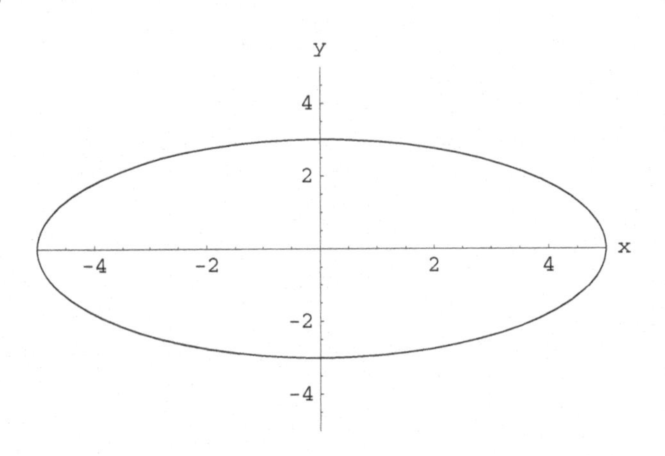

1.3 The Dot Product

3. $(-1,0,7)\cdot(2,4,-6) = -1(2)+0(4)+7(-6) = -44$, $\|(-1,0,7)\| = \sqrt{(-1)^2+0^2+7^2} = \sqrt{50} = 5\sqrt{2}$, and $\|(2,4,-6)\| = \sqrt{2^2+4^2+(-6)^2} = \sqrt{56} = 2\sqrt{14}$.

7. $\theta = \cos^{-1}\left(\frac{(\sqrt{3}\mathbf{i}+\mathbf{j})\cdot(-\sqrt{3}\mathbf{i}+\mathbf{j})}{\|(\sqrt{3}\mathbf{i}+\mathbf{j})\|\ \|-\sqrt{3}\mathbf{i}+\mathbf{j}\|}\right) = \cos^{-1}\left(\frac{-3+1}{(2)(2)}\right) = \cos^{-1}\left(\frac{-1}{2}\right) = \frac{2\pi}{3}$.

13. $\text{proj}_{\frac{\mathbf{i}+\mathbf{j}}{\sqrt{2}}}(2\mathbf{i}+3\mathbf{j}-\mathbf{k}) = \left(\frac{\left(\frac{\mathbf{i}+\mathbf{j}}{\sqrt{2}}\right)\cdot(2\mathbf{i}+3\mathbf{j}-\mathbf{k})}{\left(\frac{\mathbf{i}+\mathbf{j}}{\sqrt{2}}\right)\cdot\left(\frac{\mathbf{i}+\mathbf{j}}{\sqrt{2}}\right)}\right)\left(\frac{\mathbf{i}+\mathbf{j}}{\sqrt{2}}\right) = \frac{\frac{1}{\sqrt{2}}(2+3)}{\frac{1+1}{2}}\frac{(1,1,0)}{\sqrt{2}} = \left(\frac{5}{2},\frac{5}{2},0\right)$.

17. We just divide the vector by its length: $\frac{2\mathbf{i}-\mathbf{j}+\mathbf{k}}{\|2\mathbf{i}-\mathbf{j}+\mathbf{k}\|} = \frac{1}{\sqrt{6}}(2,-1,1)$.

21. We have two cases to consider.
If either of the projections is zero: $\text{proj}_{\mathbf{a}}\mathbf{b} = \mathbf{0} \Leftrightarrow \mathbf{a}\cdot\mathbf{b} = 0 \Leftrightarrow \text{proj}_{\mathbf{b}}\mathbf{a} = \mathbf{0}$.
If neither of the projections is zero, then the directions must be the same. This means that **a** must be a multiple of **b**. Let $\mathbf{a} = c\mathbf{b}$, then on the one hand

$$\text{proj}_{\mathbf{a}}\mathbf{b} = \text{proj}_{c\mathbf{b}}\mathbf{b} = \frac{c\mathbf{b}\cdot\mathbf{b}}{c\mathbf{b}\cdot c\mathbf{b}}c\mathbf{b} = \mathbf{b}.$$

On the other hand

$$\text{proj}_{\mathbf{b}}\mathbf{a} = \text{proj}_{\mathbf{b}}c\mathbf{b} = \frac{\mathbf{b}\cdot c\mathbf{b}}{\mathbf{b}\cdot\mathbf{b}}\mathbf{b} = c\mathbf{b}.$$

These are equal only when $c = 1$.
In other words, $\text{proj}_{\mathbf{a}}\mathbf{b} = \text{proj}_{\mathbf{b}}\mathbf{a}$ when $\mathbf{a}\cdot\mathbf{b} = 0$ or when $\mathbf{a} = \mathbf{b}$.

23. We have $\|k\mathbf{a}\| = \sqrt{k\mathbf{a}\cdot k\mathbf{a}} = \sqrt{k^2(\mathbf{a}\cdot\mathbf{a})} = \sqrt{k^2}\sqrt{\mathbf{a}\cdot\mathbf{a}} = |k|\ \|\mathbf{a}\|$.

27. To move the bananas, one must exert an *upward* force of 500 lb. Such a force makes an angle of 60° with the ramp, and it is the ramp that gives the direction of displacement. Thus the amount of work done is

$$\|\mathbf{F}\|\ \|\overrightarrow{PQ}\|\cos 60^\circ = 500\cdot 40\cdot\tfrac{1}{2} = 10,000 \text{ ft-lb}.$$

31. Consider the figure:

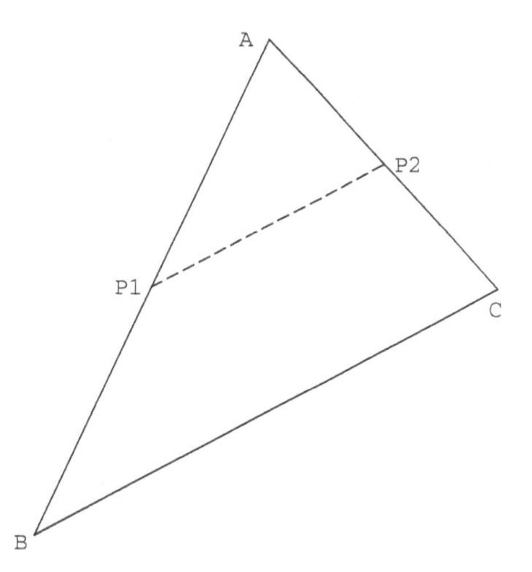

If P_1 is the point on $\overline{AB}$ located r times the distance from A to B, then the vector $\overrightarrow{AP_1} = r\overrightarrow{AB}$. Similarly, since P_2 is the point on $\overline{AC}$ located r times the distance from A to C, then the vector $\overrightarrow{AP_2} = r\overrightarrow{AC}$. So now we can look at the line segment $\overline{P_1P_2}$ using vectors.

$$\overrightarrow{P_1P_2} = \overrightarrow{AP_2} - \overrightarrow{AP_1} = r\overrightarrow{AC} - r\overrightarrow{AB} = r(\overrightarrow{AC} - \overrightarrow{AB}) = r\overrightarrow{BC}.$$

The two conclusions now follow. Because $\overrightarrow{P_1P_2}$ is a scalar multiple of $\overrightarrow{BC}$, they are parallel. Also the positive scalar r pulls out of the norm so $\|\overrightarrow{P_1P_2}\| = \|r\overrightarrow{BC}\| = r\|\overrightarrow{BC}\|$.

35. **(a)** Let's start with the two circles with centers at W_1 and W_2. Assume that in addition to their intersection at point O that they also intersect at point C as shown below.

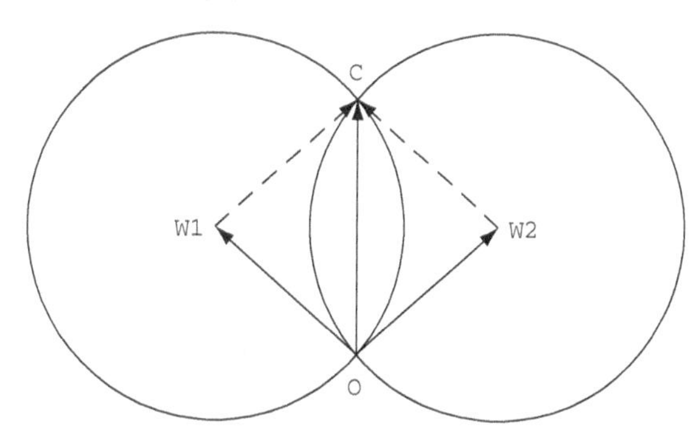

The polygon OW_1CW_2 is a parallelogram. In fact, because all sides are equal, it is a rhombus. We can, therefore, write the vector $\mathbf{c} = \overrightarrow{OC} = \overrightarrow{OW_1} + \overrightarrow{OW_2} = \mathbf{w}_1 + \mathbf{w}_2$. Similarly, we can write $\mathbf{b} = \mathbf{w}_1 + \mathbf{w}_3$ and $\mathbf{a} = \mathbf{w}_2 + \mathbf{w}_3$.

(b) Let's use the results of part (a) together with the hint. We need to show that the distance from each of the points A, B, and C to P is r. Let's show, for example, that $\|\overrightarrow{CP}\|$ is r:

$$\|\overrightarrow{CP}\| = \|\overrightarrow{OP} - \overrightarrow{OC}\| = \|(\mathbf{w}_1 + \mathbf{w}_2 + \mathbf{w}_3) - (\mathbf{w}_1 + \mathbf{w}_2)\| = \|\mathbf{w}_3\| = r.$$

The arguments for the other two points are analogous.

(c) What we really need to show is that each of the lines passing through O and one of the points A, B, or C is perpendicular to the line containing the two other points. Using vectors we will show that $\overrightarrow{OA} \perp \overrightarrow{BC}$, $\overrightarrow{OB} \perp \overrightarrow{AC}$, and $\overrightarrow{OC} \perp \overrightarrow{AB}$ by showing their dot products are 0. It's enough to show this for one of them: $\overrightarrow{OA} \cdot \overrightarrow{BC} = (\mathbf{w}_2 + \mathbf{w}_3) \cdot ((\mathbf{w}_1 + \mathbf{w}_2) - (\mathbf{w}_1 + \mathbf{w}_3)) = (\mathbf{w}_2 + \mathbf{w}_3) \cdot (\mathbf{w}_2 - \mathbf{w}_3) = \mathbf{w}_2 \cdot \mathbf{w}_2 + \mathbf{w}_3 \cdot \mathbf{w}_2 - \mathbf{w}_2 \cdot \mathbf{w}_3 - \mathbf{w}_3 \cdot \mathbf{w}_3 = r^2 - r^2 = 0$.

1.4 The Cross Product

3. Use Definition 4.2: $(1)(2)(3) + (3)(7)(-1) + (5)(0)(0) - (5)(2)(-1) - (1)(7)(0) - (3)(0)(3) = -5$.

7. Note that these two vectors form a basis for the xy-plane so the cross product will be a vector parallel to (0, 0, 1). Just using formula (3):

$$(\mathbf{i}+\mathbf{j}) \times (-3\mathbf{i}+2\mathbf{j}) = \begin{vmatrix} \mathbf{i} & \mathbf{j} & \mathbf{k} \\ 1 & 1 & 0 \\ -3 & 2 & 0 \end{vmatrix} = \begin{vmatrix} 1 & 0 \\ 2 & 0 \end{vmatrix}\mathbf{i} - \begin{vmatrix} 1 & 0 \\ -3 & 0 \end{vmatrix}\mathbf{j} + \begin{vmatrix} 1 & 1 \\ -3 & 2 \end{vmatrix}\mathbf{k} = 5\mathbf{k} = (0,0,5).$$

11. This is tricky, as the points are not given in order. The figure on the left shows the sides connected in the order that the points are given.

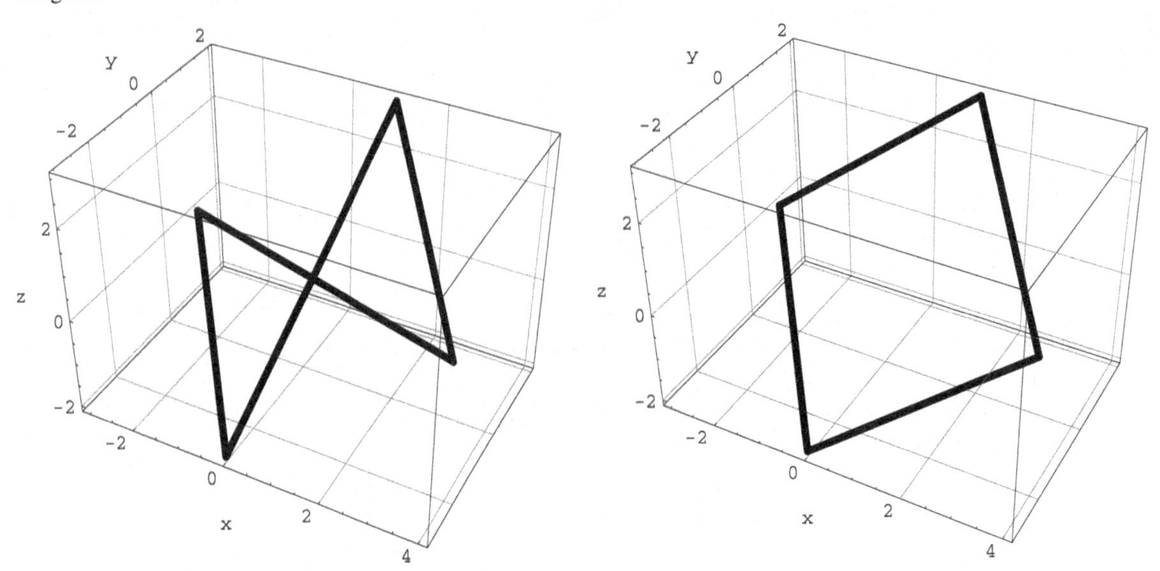

As the figure on the right shows, if you take the first side to be the side that joins the points (1, 2, 3) and $(4, -2, 1)$ then the next side is the side that joins $(4, -2, 1)$ and $(0, -3, -2)$. We will again calculate the length of the cross product of the displacement vectors. So the area of the parallelogram will be the length of

$$(0-4, -3-(-2), -2-1) \times (1-4, 2-(-2), 3-1) = (-4,-1,-3) \times (-3,4,2) = (10,17,-19).$$

The length of $(10, 17, -19)$ is $\sqrt{10^2+17^2+(-19)^2} = \sqrt{750} = 5\sqrt{30}$.

15. Take half of the length of the cross product. $(1/2)\|(1,-2,6) \times (4,3,-1)\| = (1/2)\|(-16,25,11)\| = \sqrt{1002}/2$.

19. You need to figure out a useful ordering of the vertices. You can either plot them by hand or use a computer package to help or you can make some observations about them. First look at the z coordinates. Two points have $z = -1$ and two have $z = 0$. These form your bottom face. Of the remaining points two have $z = 5$—these will match up with the bottom points with $z = -1$, and two have $z = 6$—these will match up with the bottom points with $z = 0$. The parallelepiped is shown below.

We'll use the highlighted edges as our three vectors **a**, **b**, and **c**. You could have based the calculation at any vertex. I have chosen $(4, 2, -1)$. The three vectors are:

$$\mathbf{a} = (0,3,0) - (4,2,-1) = (-4,1,1)$$
$$\mathbf{b} = (4,3,5) - (4,2,-1) = (0,1,6)$$
$$\mathbf{c} = (3,0,-1) - (4,2,-1) = (-1,-2,0)$$

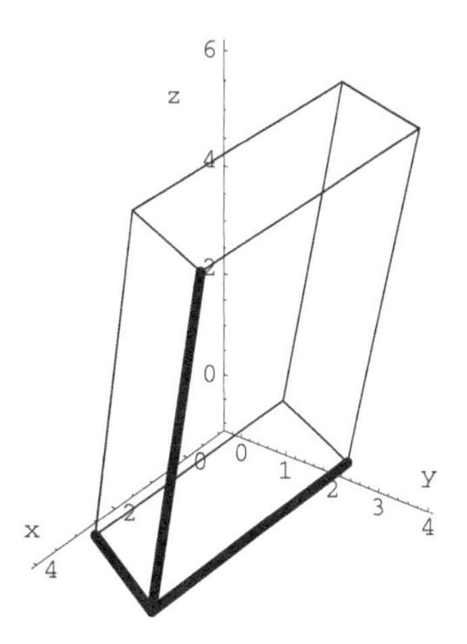

We can now calculate

$$\begin{aligned}(\mathbf{a}\times\mathbf{b})\cdot\mathbf{c} &= ((-4,1,1)\times(0,1,6))\cdot(-1,-2,0) = \begin{vmatrix} -4 & 1 & 1 \\ 0 & 1 & 6 \\ -1 & -2 & 0 \end{vmatrix} \\ &= -4\begin{vmatrix} 1 & 6 \\ -2 & 0 \end{vmatrix} - 1\begin{vmatrix} 0 & 6 \\ -1 & 0 \end{vmatrix} + 1\begin{vmatrix} 0 & 1 \\ -1 & -2 \end{vmatrix} = -4(12)-(6)+(1) = -53.\end{aligned}$$

Finally, Volume $= |(\mathbf{a}\times\mathbf{b})\cdot\mathbf{c}| = 53$.

23. (a) We have

$$\begin{aligned}\text{Area} &= \tfrac{1}{2}\|\overrightarrow{P_1P_2}\times\overrightarrow{P_1P_3}\| \\ &= \tfrac{1}{2}\|(x_2-x_1, y_2-y_1, 0)\times(x_3-x_1, y_3-y_1, 0)\|\end{aligned}$$

$$\text{Now } \overrightarrow{P_1P_2}\times\overrightarrow{P_1P_3} = \begin{vmatrix} \mathbf{i} & \mathbf{j} & \mathbf{k} \\ x_2-x_1 & y_2-y_1 & 0 \\ x_3-x_1 & y_3-y_1 & 0 \end{vmatrix}$$

$$= [(x_2-x_1)(y_3-y_1)-(x_3-x_1)(y_2-y_1)]\mathbf{k}$$

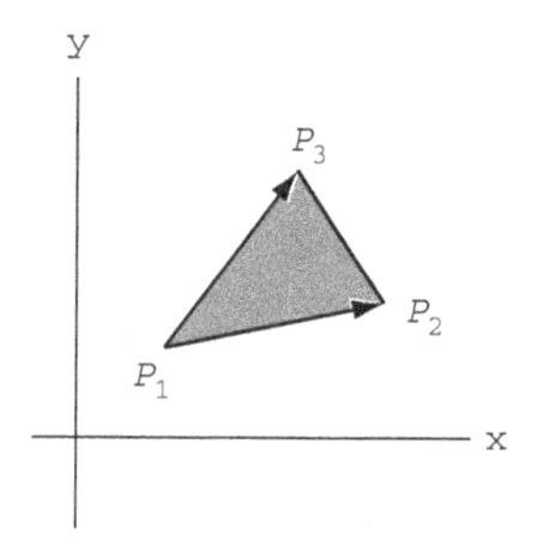

Hence the area is $\frac{1}{2}|(x_2-x_1)(y_3-y_1)-(x_3-x_1)(y_2-y_1)|$. On the other hand

$$\frac{1}{2}\begin{vmatrix} 1 & 1 & 1 \\ x_1 & x_2 & x_3 \\ y_1 & y_2 & y_3 \end{vmatrix} = \frac{1}{2}\left(\begin{vmatrix} x_2 & x_3 \\ y_2 & y_3 \end{vmatrix} - \begin{vmatrix} x_1 & x_3 \\ y_1 & y_3 \end{vmatrix} + \begin{vmatrix} x_1 & x_2 \\ y_1 & y_2 \end{vmatrix}\right).$$

Expanding and taking absolute value, we obtain

$$\tfrac{1}{2}|x_2y_3 - x_3y_2 + x_3y_1 - x_1y_3 + x_1y_2 - x_2y_1|.$$

From here, its easy to see that this agrees with the formula above.

(b) We compute the absolute value of $\frac{1}{2}\begin{vmatrix} 1 & 1 & 1 \\ 1 & 2 & -4 \\ 2 & 3 & -4 \end{vmatrix} = \frac{1}{2}(-8-8+3-4+12+4) = \frac{1}{2}(-1) = -\frac{1}{2}$.

Thus the area is $\left|-\frac{1}{2}\right| = \frac{1}{2}$.

27. Exercise 25(f) shows us that $(\mathbf{a} \times \mathbf{b}) \times \mathbf{c}$ is in the plane determined by $\mathbf{a}$ and $\mathbf{b}$ and so we expect the solution to be of the form $k_1\mathbf{a} + k_2\mathbf{b}$ for scalars k_1 and k_2.

Using formula (3) from the text for $\mathbf{a} \times \mathbf{b}$:

$$
\begin{aligned}
(\mathbf{a} \times \mathbf{b}) \times \mathbf{c} &= \begin{vmatrix} \mathbf{i} & \mathbf{j} & \mathbf{k} \\ \begin{vmatrix} a_2 & a_3 \\ b_2 & b_3 \end{vmatrix} & -\begin{vmatrix} a_1 & a_3 \\ b_1 & b_3 \end{vmatrix} & \begin{vmatrix} a_1 & a_2 \\ b_1 & b_2 \end{vmatrix} \\ c_1 & c_2 & c_3 \end{vmatrix} \\
&= \left(-\begin{vmatrix} a_1 & a_3 \\ b_1 & b_3 \end{vmatrix} c_3 - \begin{vmatrix} a_1 & a_2 \\ b_1 & b_2 \end{vmatrix} c_2\right)\mathbf{i} - \left(\begin{vmatrix} a_2 & a_3 \\ b_2 & b_3 \end{vmatrix} c_3 - \begin{vmatrix} a_1 & a_2 \\ b_1 & b_2 \end{vmatrix} c_1\right)\mathbf{j} \\
&\quad + \left(\begin{vmatrix} a_2 & a_3 \\ b_2 & b_3 \end{vmatrix} c_2 + \begin{vmatrix} a_1 & a_3 \\ b_1 & b_3 \end{vmatrix} c_1\right)\mathbf{k}
\end{aligned}
$$

Look first at the coefficient of $\mathbf{i}$: $-a_1b_3c_3 + a_3b_1c_3 - a_1b_2c_2 + a_2b_1c_2$. If we add and subtract $a_1b_1c_1$ and regroup we have: $b_1(a_1c_1 + a_2c_2 + a_3c_3) - a_1(b_1c_1 + b_2c_2 + b_3c_3) = b_1(\mathbf{a} \cdot \mathbf{c}) - a_1(\mathbf{b} \cdot \mathbf{c})$. Similarly for the coefficient of $\mathbf{j}$. Expand then add and subtract $a_2b_2b_3$ and regroup to get $b_2(\mathbf{a} \cdot \mathbf{c}) - a_2(\mathbf{b} \cdot \mathbf{c})$. Finally for the coefficient of $\mathbf{k}$, expand then add and subtract $a_3b_3c_3$ and regroup to obtain $b_3(\mathbf{a} \cdot \mathbf{c}) - a_3(\mathbf{b} \cdot \mathbf{c})$. This shows that $(\mathbf{a} \times \mathbf{b}) \times \mathbf{c} = (\mathbf{a} \cdot \mathbf{c})\mathbf{b} - (\mathbf{b} \cdot \mathbf{c})\mathbf{a}$.

Now here's a version of Exercise 27 worked on *Mathematica*. First you enter the following to define the vectors **a**, **b**, and **c**.

$$
\begin{aligned}
a &= \{a1, a2, a3\} \\
b &= \{b1, b2, b3\} \\
c &= \{c1, c2, c3\}
\end{aligned}
$$

The reply from *Mathematica* is an echo of your input for **c**. Let's begin by calculating the cross product. You can either select the cross product operator from the typesetting palette or you can type the escape key followed by "cross" followed by the escape key. *Mathematica* should reformat this key sequence as $\times$ and you should be able to enter

$$(a \times b) \times c.$$

Mathematica will respond with the calculated cross product

$$
\begin{aligned}
&\{a2b1c2 - a1b2c2 + a3b1c3 - a1b3c3, \\
&-a2b1c1 + a1b2c1 + a3b2c3 - a2b3c3, \\
&-a3b1c1 + a1b3c1 - a3b2c2 + a2b3c2\}.
\end{aligned}
$$

Now you can check the other expression. Use a period for the dot in the dot product.

$$(a.c)b - (b.c)\ a$$

Mathematica will immediately respond

$$
\begin{aligned}
&\{b1(a1c1 + a2c2 + a3c3) - a1(b1c1 + b2c2 + b3c3), \\
&b2(a1c1 + a2c2 + a3c3) - a2(b1c1 + b2c2 + b3c3), \\
&b3(a1c1 + a2c2 + a3c3) - a3(b1c1 + b2c2 + b3c3)\}
\end{aligned}
$$

This certainly looks different from the previous expression. Before giving up hope, note that this one has been factored and the earlier one has not. You can expand this by using the command

$$\text{Expand}[(a.c)b - (b.c)a]$$

or use *Mathematica*'s command % to refer to the previous entry and just type

$$\text{Expand}[\%].$$

This still might not look familiar. So take a look at

$$(a \times b) \times c - [(a.c)b - (b.c)a].$$

If this *still* isn't what you are looking for, simplify it with the command

$$\text{Simplify}[\%]$$

and *Mathematica* will respond

$$\{0, 0, 0\}.$$

31. If you are using a computer algebra system, you may not notice that this is *exactly* the same problem as Exercise 27. Just replace $\mathbf{c}$ with $(\mathbf{c} \times \mathbf{d})$ on both sides of the equation in Exercise 27 to obtain the result here.

35. **(a)** Here the length of $\mathbf{a}$ is 1 foot, the force $\mathbf{F} = 40$ pounds and angle $\theta = 120$ degrees. So

$$\textbf{Torque} = (1)(40)\sin 120^\circ = 40\left(\frac{\sqrt{3}}{2}\right) = 20\sqrt{3} \text{ foot-pounds.}$$

(b) Here all that has changed is that $\|\mathbf{a}\|$ is 1.5 feet, so

$$\textbf{Torque} = (3/2)(40)\sin 120^\circ = 60\left(\frac{\sqrt{3}}{2}\right) = 30\sqrt{3} \text{ foot-pounds.}$$

39. Archie's actual experience isn't important in solving this problem; he could have ridden closer to the center. Since we are only interested in comparing Archie's experience with Annie's, it turns out that their difference would be the same so long as the difference in their distance from the center remained at 2 inches. The difference in speed is $(331/3)(2\pi)(6) - (331/3)(2\pi)(4) = (331/3)(2\pi)(2) = 4\pi(331/3) = 1331/3\pi = 400\pi/3$ in/ min.

1.5 Equations for Planes; Distance Problems

3. We first need to find a vector normal to the plane, so we take the cross product of two displacement vectors:

$$(3-2, -1-0, 2-5) \times (1-2, -2-0, 4-5) = (1, -1, -3) \times (-1, -2, -1) = (-5, 4, -3).$$

Now we can apply formula (2) using any of the three points:

$$-5(x-3) + 4(y+1) - 3(z-2) = 0 \iff -5x + 4y - 3z = -25.$$

7. We may take the normal to the plane to be the same as a normal to the given plane; thus we may let $\mathbf{n} = \mathbf{i} - \mathbf{j} + 7\mathbf{k}$. Hence an equation for the desired plane is

$$1(x+2) - 1(y-0) + 7(z-1) = 0 \iff x - y + 7z = 5.$$

13. The line shared by two planes will be orthogonal each of their normal vectors. First, calculate: $(1, 2, -3) \times (5, 5, -1) = (13, -14, -5)$. Now find a point on the line by setting $z = 0$ and solving the two equations

$$\begin{cases} x + 2y = 5 \\ 5x + 5y = 1 \end{cases}$$

to get $x = -23/5$ and $y = 24/5$. The equation of the line is $\mathbf{r}(t) = (-23/5, 24/5, 0) + (13, -14, -5)t$, or in parametric form:

$$\begin{cases} x &= 13t - \frac{23}{5} \\ y &= -14t + \frac{24}{5} \\ z &= -5t. \end{cases}$$

17. This is a direct application of formula (10):

$$\mathbf{x}(s,t) = s\mathbf{a} + t\mathbf{b} + \mathbf{c} = s(2, -3, 1) + t(1, 0, -5) + (-1, 2, 7).$$

In parametric form this is:

$$\begin{cases} x = 2s + t - 1 \\ y = -3s + 2 \\ z = s - 5t + 7 \end{cases}$$

21. The equation of the line $\mathbf{r}(t) = (-5, 10, 9) + t(3, -3, 2)$ immediately gives us one of the two vectors $\mathbf{a} = (3, -3, 2)$. The displacement vector from a point on the line to our given point gives us the vector $\mathbf{b} = (-5, 10, 9) - (-2, 4, 7) = (-3, 6, 2)$. So our equations are:

$$\mathbf{x}(s,t) = s(3,-3,2) + t(-3,6,2) + (-5,10,9) \quad \text{or} \quad \begin{cases} x = 3s - 3t - 5 \\ y = -3s + 6t + 10 \\ z = 2s + 2t + 9. \end{cases}$$

23. We combine the parametric equations into the single equation:

$$\mathbf{x}(s,t) = s(3,4,1) + t(-1,1,5) + (2,0,3).$$

Use the cross product to find the normal vector to the plane:

$$\mathbf{n} = (3,4,1) \times (-1,1,5) = (19,-16,7).$$

So the equation of the plane is:

$$19(x-2) - 16y + 7(z-3) = 0 \quad \text{or} \quad 19x - 16y + 7z = 59.$$

27. Use Example 9 and for two points $B_1 = (-1, 3, 5)$ on l_1 and $B_2 = (0, 3, 4)$ on l_2 calculate $\overrightarrow{B_1B_2} = (1, 0, -1)$. To find the vector $\mathbf{n}$, calculate the cross product $\mathbf{n} = (8, -1, 0) \times (0, 3, 1) = (-1, -8, 24)$.

$$\begin{aligned}\text{proj}_{\mathbf{n}}\overrightarrow{B_1B_2} &= \left(\frac{\mathbf{n}\cdot\overrightarrow{B_1B_2}}{\mathbf{n}\cdot\mathbf{n}}\right)\mathbf{n} = \left(\frac{(-1,-8,24)\cdot(1,0,-1)}{(-1,-8,24)\cdot(-1,-8,24)}\right)(-1,-8,24) \\ &= -\frac{25}{641}(-1,-8,24).\end{aligned}$$

Finally $\left\|-\frac{25}{641}(-1,-8,24)\right\| = \frac{25}{\sqrt{641}}$.

31. These planes are parallel so we can use Example 8. The point $P_1 = (1, 0, 0)$ is on plane one and the point $P_2 = (8, 0, 0)$ is on plane two. We project the displacement vector $\overrightarrow{P_1P_2} = (7, 0, 0)$ onto the normal direction $\mathbf{n} = (1, -3, 2)$:

$$\text{proj}_{\mathbf{n}}\overrightarrow{P_1P_2} = \left(\frac{(7,0,0)\cdot(1,-3,2)}{(1,-3,2)\cdot(1,-3,2)}\right)(1,-3,2) = \frac{7}{14}(1,-3,2) = \frac{1}{2}(1,-3,2).$$

So the distance is $\|\text{proj}_{\mathbf{n}}\overrightarrow{P_1P_2}\| = \sqrt{14}/2$.

33. As in Exercises 27 and 28, we'll choose a point $P_1 = (D_1/A, 0, 0)$ on plane one and $P_2 = (D_2/A, 0, 0)$ on plane two. The displacement vector is

$$\overrightarrow{P_1P_2} = \left(\frac{D_2 - D_1}{A}, 0, 0\right).$$

A vector normal to the plane is $\mathbf{n} = (A, B, C)$.

$$\text{proj}_{\mathbf{n}}\overrightarrow{P_1P_2} = \left(\frac{(\frac{D_2-D_1}{A},0,0)\cdot(A,B,C)}{(A,B,C)\cdot(A,B,C)}\right)(A,B,C) = \frac{D_2-D_1}{A^2+B^2+C^2}(A,B,C).$$

The distance between the two planes is:

$$\|\text{proj}_{\mathbf{n}}\overrightarrow{P_1P_2}\| = \frac{|D_2-D_1|}{A^2+B^2+C^2}\|(A,B,C)\| = \frac{|D_2-D_1|}{\sqrt{A^2+B^2+C^2}}.$$

35. This exercise follows immediately from Exercise 33 (and can be very difficult without it). Here $A = 1, B = 3, C = -5$ and $D_1 = 2$. The equation in Exercise 33 becomes:

$$3 = \frac{|2-D_2|}{\sqrt{1^2+3^2+(-5)^2}} = \frac{|2-D_2|}{\sqrt{35}}.$$

So

$$3\sqrt{35} = |2-D_2| \quad \text{or} \quad 2 - D_2 = \pm 3\sqrt{35}.$$

So $D_2 = 2 \pm 3\sqrt{35}$ and the equations of the two planes are:

$$x + 3y - 5z = 2 \pm 3\sqrt{35}.$$

39. By letting $t = 0$ in each vector parametric equation, we obtain $\mathbf{b}_1, \mathbf{b}_2$ as position vectors of points B_1, B_2 on the respective lines. Hence $\overrightarrow{B_1B_2} = \mathbf{b}_2 - \mathbf{b}_1$. A vector $\mathbf{n}$ perpendicular to both lines is given by $\mathbf{n} = \mathbf{a}_1 \times \mathbf{a}_2$. Thus

$$\begin{aligned} D = \|\text{proj}_{\mathbf{n}} \overrightarrow{B_1B_2}\| &= \frac{|\mathbf{n} \cdot \overrightarrow{B_1B_2}|}{\|\mathbf{n}\|^2}\|\mathbf{n}\| = \frac{|\mathbf{n} \cdot \overrightarrow{B_1B_2}|}{\|\mathbf{n}\|} \\ &= \frac{|(\mathbf{a}_1 \times \mathbf{a}_2) \cdot (\mathbf{b}_2 - \mathbf{b}_1)|}{\|\mathbf{a}_1 \times \mathbf{a}_2\|}. \end{aligned}$$

1.6 Some n-dimensional Geometry

1. **(a)** $(1, 2, 3, \ldots, n) = (1, 0, 0, \ldots, 0) + 2(0, 1, 0, 0, \ldots, 0) + \cdots + n(0, 0, 0, \ldots, 0, 1) = \mathbf{e}_1 + 2\mathbf{e}_2 + 3\mathbf{e}_3 + \cdots + n\mathbf{e}_n$.
(b) $(1, 0, -1, 1, 0, -1, \ldots, 1, 0, -1) = \mathbf{e}_1 - \mathbf{e}_3 + \mathbf{e}_4 - \mathbf{e}_6 + \mathbf{e}_7 - \mathbf{e}_9 + \cdots + \mathbf{e}_{n-2} - \mathbf{e}_n$.

7. First we calculate

$$\|\mathbf{a}\| = \sqrt{1^2 + 2^2 + 3^2 + \cdots + n^2} = \sqrt{\frac{n(n+1)(2n+1)}{6}}$$

$$\|\mathbf{b}\| = \underbrace{\sqrt{1^2 + 1^2 + \cdots + 1^2}}_{n \text{ times}} = \sqrt{n}, \text{ and}$$

$$|\mathbf{a} \cdot \mathbf{b}| = 1 + 2 + 3 + \cdots + n = \frac{n(n+1)}{2}$$

So

$$\|\mathbf{a}\| \, \|\mathbf{b}\| = \left(\sqrt{\frac{n(n+1)(2n+1)}{6}}\right)(\sqrt{n}) = \left(\frac{n}{2}\right)\left(\sqrt{\frac{2(n+1)(2n+1)}{3}}\right).$$

For $n = 1$, $\sqrt{\frac{2(n+1)(2n+1)}{3}} = 2 = n + 1$.
For $n = 2$, $\sqrt{\frac{2(n+1)(2n+1)}{3}} = \sqrt{10} \geq 3 = n + 1$.
For $n \geq 3$,

$$\left(\frac{n}{2}\right)\left(\sqrt{\frac{2(n+1)(2n+1)}{3}}\right) \geq \left(\frac{n}{2}\right)\frac{2n+1}{\sqrt{3}} \geq \left(\frac{n}{2}\right)(n+1) = |\mathbf{a} \cdot \mathbf{b}|.$$

11. We have

$$\begin{aligned} \|\mathbf{a} + \mathbf{b}\| = \|\mathbf{a} - \mathbf{b}\| &\Rightarrow \|\mathbf{a} + \mathbf{b}\|^2 = \|\mathbf{a} - \mathbf{b}\|^2 \\ &\Rightarrow (\mathbf{a} + \mathbf{b}) \cdot (\mathbf{a} + \mathbf{b}) = (\mathbf{a} - \mathbf{b}) \cdot (\mathbf{a} - \mathbf{b}). \end{aligned}$$

Expand to find

$$\begin{aligned} \mathbf{a} \cdot \mathbf{a} + 2\mathbf{a} \cdot \mathbf{b} + \mathbf{b} \cdot \mathbf{b} &= \mathbf{a} \cdot \mathbf{a} - 2\mathbf{a} \cdot \mathbf{b} + \mathbf{b} \cdot \mathbf{b} \\ &\Rightarrow 4\mathbf{a} \cdot \mathbf{b} = 0, \end{aligned}$$

so **a** and **b** are orthogonal.

15. **(a)** We have

$$\mathbf{p} = (200, 250, 300, 375, 450, 500)$$
$$\text{Total cost} = \mathbf{p} \cdot \mathbf{x} = 200x_1 + 250x_2 + 300x_3 + 375x_4 + 450x_5 + 500x_6$$

(b) With **p** as in part (a), the customer can afford commodity bundles **x** in the set

$$\{\mathbf{x} \in \mathbf{R}^6 | \mathbf{p} \cdot \mathbf{x} \leq 100{,}000\}.$$

The budget hyperplane is $\mathbf{p} \cdot \mathbf{x} = 100{,}000$ or $200x_1 + 250x_2 + 300x_3 + 375x_4 + 450x_5 + 500x_6 = 100{,}000$.

19.

$$B^T D = \begin{bmatrix} -4 & 0 \\ 9 & 3 \\ 5 & 0 \end{bmatrix} \begin{bmatrix} 1 & 0 \\ 2 & -3 \end{bmatrix}$$
$$= \begin{bmatrix} -4(1)+0(2) & -4(0)+0(-3) \\ 9(1)+3(2) & 9(0)+3(-3) \\ 5(1)+0(2) & 5(0)+0(-3) \end{bmatrix}$$
$$= \begin{bmatrix} -4 & 0 \\ 15 & -9 \\ 5 & 0 \end{bmatrix}$$

23. This is similar to Exercise 22. Either we could expand along the last row of each matrix at each step or we could expand along the first column at each step. It is easier to keep track of signs if we choose this second approach. We again find that the determinant is the product of the diagonal elements.

$$\begin{vmatrix} 5 & -1 & 0 & 8 & 11 \\ 0 & 2 & 1 & 9 & 7 \\ 0 & 0 & 4 & -3 & 5 \\ 0 & 0 & 0 & 2 & 1 \\ 0 & 0 & 0 & 0 & -3 \end{vmatrix} = 5 \begin{vmatrix} 2 & 1 & 9 & 7 \\ 0 & 4 & -3 & 5 \\ 0 & 0 & 2 & 1 \\ 0 & 0 & 0 & -3 \end{vmatrix}$$
$$= (5)(2) \begin{vmatrix} 4 & -3 & 5 \\ 0 & 2 & 1 \\ 0 & 0 & -3 \end{vmatrix}$$
$$= (5)(2)(4) \begin{vmatrix} 2 & 1 \\ 0 & -3 \end{vmatrix}$$
$$= (5)(2)(4)(2)(-3) = -240.$$

25. **(a)** **A lower triangular** matrix is an $n \times n$ matrix whose entries above the main diagonal are all zero. For example the matrix in Exercise 22 is lower triangular.

(b) If we expand the determinant of an upper triangular matrix along its first column we get:

$$|A| = (-1)^{1+1} a_{11} |A_{11}| + (-1)^{2+1} a_{21} |A_{21}| + \cdots + (-1)^{n+1} a_{n1} |A_{n1}|$$
$$= (-1)^{1+1} (a_{11}) |A_{11}| + (-1)^{2+1} (0) |A_{2j}| + \cdots + (-1)^{n+1} (0) |A_{nj}| = (a_{11}) |A_{11}|.$$

Looking back on what we have found: The determinant of an upper triangular matrix is equal to the term in the upper left position multiplied by the determinant of the matrix that's left when the top most row and left most column are removed. Each time we remove the top row and left column we are left with an upper triangular matrix of one dimension lower. Repeat the process n times and it is clear that

$$|A| = a_{11}|A_{11}| = a_{11}(a_{22}|(A_{11})_{11}|) = \cdots = a_{11}a_{22}a_{33}\cdots a_{nn}.$$

29. This is a pretty cool fact. If *AB* and *BA* both exist, these two matrices may not be equal. It doesn't matter. They still have the same determinant. The proof is straightforward: $\det(AB) = (\det\ A)(\det B) = (\det\ B)(\det\ A) = \det\ (BA)$.

33. Using the hint, assume that A has two inverses B and C. Then

$$B = BI = B(AC) = (BA)C = IC = C.$$

37. If $A = \begin{bmatrix} 2 & 1 & 1 \\ 0 & 2 & 4 \\ 1 & 0 & 3 \end{bmatrix}$, then $\det A = 12 + 4 - 2 = 14$, so the formula gives

$$A^{-1} = \frac{1}{14}\begin{bmatrix} \begin{vmatrix} 2 & 4 \\ 0 & 3 \end{vmatrix} & -\begin{vmatrix} 1 & 1 \\ 0 & 3 \end{vmatrix} & \begin{vmatrix} 1 & 1 \\ 2 & 4 \end{vmatrix} \\ -\begin{vmatrix} 0 & 4 \\ 1 & 3 \end{vmatrix} & \begin{vmatrix} 2 & 1 \\ 1 & 3 \end{vmatrix} & -\begin{vmatrix} 2 & 1 \\ 0 & 4 \end{vmatrix} \\ \begin{vmatrix} 0 & 2 \\ 1 & 0 \end{vmatrix} & -\begin{vmatrix} 2 & 1 \\ 1 & 0 \end{vmatrix} & \begin{vmatrix} 2 & 1 \\ 0 & 2 \end{vmatrix} \end{bmatrix}$$

$$= \frac{1}{14}\begin{bmatrix} 6 & -3 & 2 \\ 4 & 5 & -8 \\ -2 & 1 & 4 \end{bmatrix} = \begin{bmatrix} \frac{3}{7} & -\frac{3}{14} & \frac{1}{7} \\ \frac{2}{7} & \frac{5}{14} & -\frac{4}{7} \\ -\frac{1}{7} & \frac{1}{14} & \frac{2}{7} \end{bmatrix}$$

41. This follows immediately from part (d) of Exercise 28. For $1 \le i \le n-1$,

$$\mathbf{a}_i \cdot (\mathbf{a}_1 \times \cdots \times \mathbf{a}_i \times \cdots \times \mathbf{a}_{n-1}) = \begin{vmatrix} a_{i1} & a_{i2} & \cdots & a_{in} \\ a_{11} & a_{12} & \cdots & a_{1n} \\ & & \ddots & \\ \vdots & \vdots & & \vdots \\ a_{(n-1)1} & a_{(n-1)2} & \cdots & a_{(n-1)n} \end{vmatrix}.$$

Replace the first row with the difference between row 1 and row $i+1$ and you will get (by Exercise 26) a matrix with the same determinant, namely:

$$\begin{vmatrix} 0 & 0 & \cdots & 0 \\ a_{11} & a_{12} & \cdots & a_{1n} \\ & & \ddots & \\ \vdots & \vdots & & \vdots \\ a_{(n-1)1} & a_{(n-1)2} & \cdots & a_{(n-1)n} \end{vmatrix} = 0.$$

Therefore $\mathbf{b}$ is orthogonal to $\mathbf{a}_i$ for $1 \le i \le n-1$.

1.7 New Coordinate Systems

1. Use equations (1) $x = r\cos\theta$ and $y = r\sin\theta$. $x = \sqrt{2}\cos\pi/4 = (\sqrt{2})(\sqrt{2}/2) = 1$, and $y = \sqrt{2}\sin\pi/4 = 1$. The rectangular coordinates are (1, 1).

5. Use equations (2) $r^2 = x^2 + y^2$ and $\tan\theta = y/x$. $r^2 = (-2)^2 + 2^2 = 8$, so $r = 2\sqrt{2}$. Also, $\tan\theta = 2/(-2) = -1$. Since we are in the second quadrant the polar coordinates are $(2\sqrt{2}, 3\pi/4)$.

9. Use equations (3). $x = 1\cos 2\pi/3 = -1/2$, $y = 1\sin 2\pi/3 = \sqrt{3}/2$, and $z = -2$. The rectangular coordinates are $(-1/2, \sqrt{3}/2, -2)$.

13. Use equations (7) $x = \rho\sin\varphi\cos\theta$, $y = \rho\sin\varphi\sin\theta$, and $z = \rho\cos\varphi$. $x = 2(\sin\pi)(\cos\pi/4) = 2(0)(\sqrt{2}/2) = 0$, $y = 2(\sin\pi)(\sin\pi/4) = 2(0)(\sqrt{2}/2) = 0$, and $z = 2\cos\pi = 2(-1) = -2$. So the rectangular coordinates are $(0, 0, -2)$.

17. Use equations (7) $\rho^2 = x^2 + y^2 + z^2$, $\tan\varphi = \sqrt{x^2+y^2}/z$, and $\tan\theta = y/x$. $\rho^2 = (1)^2 + (-1)^2 + (\sqrt{6})^2 = 8$, so $\rho = \sqrt{8} = 2\sqrt{2}$. Also, $\tan\varphi = \sqrt{1^2 + (-1)^2}/\sqrt{6} = \sqrt{2}/\sqrt{6} = (1/2)/(\sqrt{3}/2)$, so $\varphi = \pi/6$. Finally, $\tan\theta = -1/1 = -1$, so $\theta = 7\pi/4$ (since the point, when projected onto the xy-plane is in the fourth quadrant). In spherical coordinates the point is $(2\sqrt{2}, \pi/6, 7\pi/4)$.

19. As in Example 5, θ does not appear so the surface will be circularly symmetric about the z-axis. Once we have our answer to part (a), we can just rotate it about the z-axis to generate the answer to part (b).

(a) We are slicing in the direction $\pi/2$ which puts us in the yz-plane for positive y. This means that $(r-2)^2 + z^2 = 1$ becomes $(y-2)^2 + z^2 = 1$. This is a circle of radius 1 centered at (0, 2, 0).

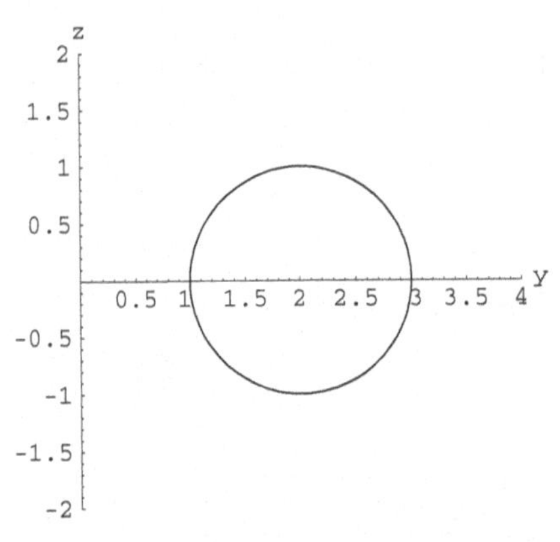

(b) As we start to rotate this about the z-axis, we get a feel for the shape being generated (see below left). In the figure above we see the result of the condition that $r \geq 0$. Without that restriction we would see two circles, each sweeping out a trail like that above. We would end up tracing our surface twice. Rotating this circle (with the restriction on r) about the z-axis, we will end up with a torus (see below right).

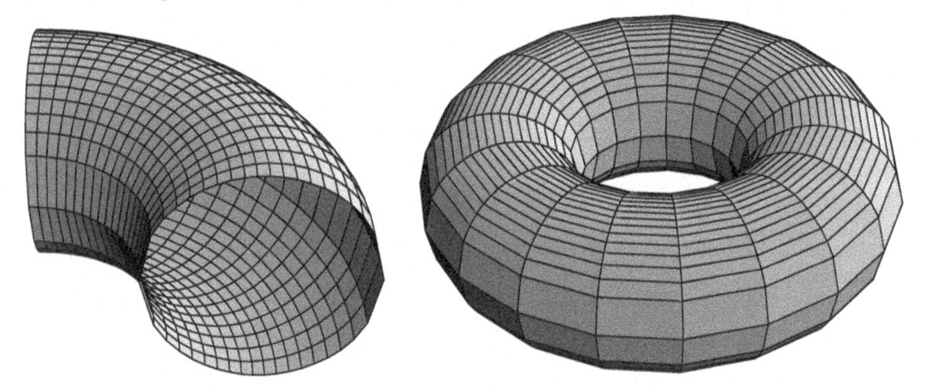

23. The equation: $\rho \sin \varphi \sin \theta = 2$ is clearly a spherical equation (it involves all three of the spherical coordinates).

- Use equation (7) to convert it to cartesian coordinates: $y = \rho \sin \varphi \sin \theta$ so the cartesian form is simply

$$y = 2.$$

  This is a vertical plane parallel to the xz-plane.

- Use equation (6) to convert to cylindrical coordinates. $\sin \theta$ stays $\sin \theta$ and $\rho \sin \varphi = r$. So the cylindrical form is

$$r \sin \theta = 2.$$

29. This solid is bound by two paraboloids.

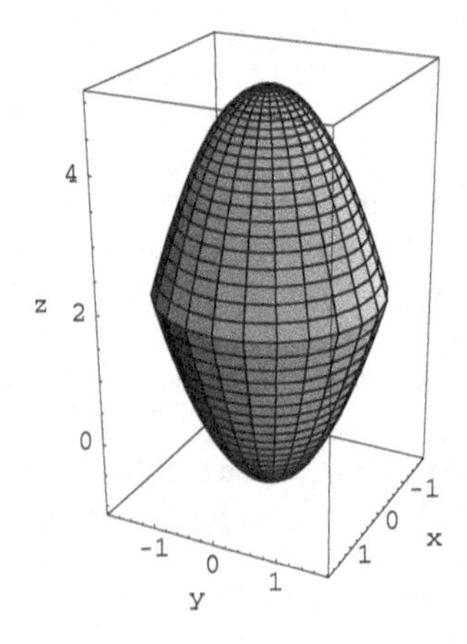

33. This looks like an ice cream cone:

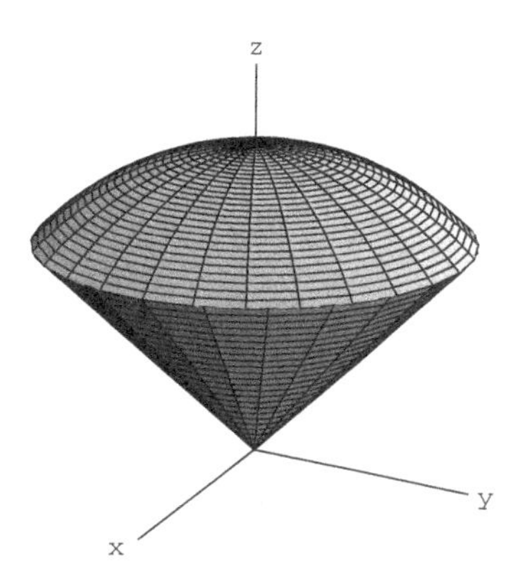

35. This is a sphere of radius 3 centered at the origin from which we've removed a sphere of radius 1 centered at $(0, 0, 1)$.

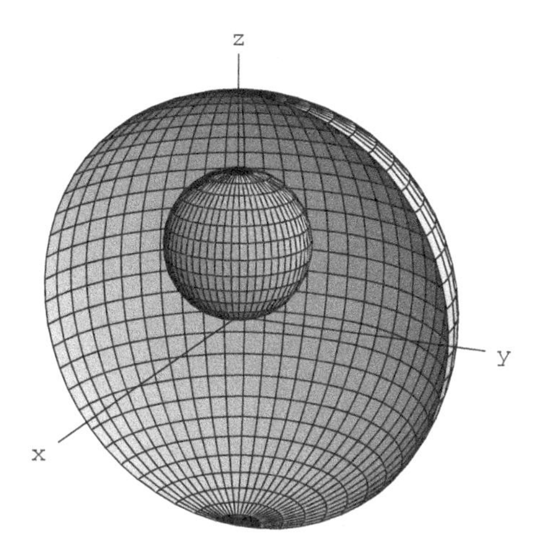

39. **(a)** We need to take six dot products. Each vector dotted with itself must be 1 and each vector dotted with any other must be 0.

$$\mathbf{e}_r \cdot \mathbf{e}_r = (\cos\theta, \sin\theta, 0) \cdot (\cos\theta, \sin\theta, 0) = \cos^2\theta + \sin^2\theta = 1.$$

$$\mathbf{e}_\theta \cdot \mathbf{e}_\theta = (-\sin\theta, \cos\theta, 0) \cdot (-\sin\theta, \cos\theta, 0) = \sin^2\theta + \cos^2\theta = 1.$$

$$\mathbf{e}_z \cdot \mathbf{e}_z = (0, 0, 1) \cdot (0, 0, 1) = 1.$$

$$\mathbf{e}_r \cdot \mathbf{e}_\theta = (\cos\theta, \sin\theta, 0) \cdot (-\sin\theta, \cos\theta, 0) = -\cos\theta\sin\theta + \sin\theta\cos\theta = 0.$$

$$\mathbf{e}_r \cdot \mathbf{e}_z = (\cos\theta, \sin\theta, 0) \cdot (0, 0, 1) = 0.$$

$$\mathbf{e}_\theta \cdot \mathbf{e}_z = (-\sin\theta, \cos\theta, 0) \cdot (0, 0, 1) = 0.$$

(b) We now do the same for the spherical basis vectors.

$$\mathbf{e}_\rho \cdot \mathbf{e}_\rho = (\sin\varphi\cos\theta, \sin\varphi\sin\theta, \cos\varphi)\cdot(\sin\varphi\cos\theta, \sin\varphi\sin\theta, \cos\varphi) = \sin^2\varphi\cos^2\theta + \sin^2\varphi\sin^2\theta + \cos^2\varphi$$
$$= \sin^2\varphi + \cos^2\varphi = 1.$$
$$\mathbf{e}_\varphi \cdot \mathbf{e}_\varphi = (\cos\varphi\cos\theta, \cos\varphi\sin\theta, -\sin\varphi)\cdot(\cos\varphi\cos\theta, \cos\varphi\sin\theta, -\sin\varphi) = \cos^2\varphi\cos^2\theta$$
$$+ \cos^2\varphi\sin^2\theta + \sin^2\varphi = \cos^2\varphi + \sin^2\varphi = 1.$$
$$\mathbf{e}_\theta \cdot \mathbf{e}_\theta = (-\sin\theta, \cos\theta, 0)\cdot(-\sin\theta, \cos\theta, 0) = \sin^2\theta + \cos^2\theta = 1.$$
$$\mathbf{e}_\rho \cdot \mathbf{e}_\varphi = (\sin\varphi\cos\theta, \sin\varphi\sin\theta, \cos\varphi)\cdot(\cos\varphi\cos\theta, \cos\varphi\sin\theta, -\sin\varphi) = \sin\varphi\cos\varphi\cos^2\theta + \sin\varphi\cos\varphi\cos^2\theta$$
$$- \sin\varphi\cos\varphi = \sin\varphi\cos\varphi - \sin\varphi\cos\varphi = 0.$$
$$\mathbf{e}_\rho \cdot \mathbf{e}_\theta = (\sin\varphi\cos\theta, \sin\varphi\sin\theta, \cos\varphi)\cdot(-\sin\theta, \cos\theta, 0) = -\sin\varphi\cos\theta\sin\theta + \sin\varphi\sin\theta\cos\theta = 0.$$
$$\mathbf{e}_\varphi \cdot \mathbf{e}_\theta = (\cos\varphi\cos\theta, \cos\varphi\sin\theta, -\sin\varphi)\cdot(-\sin\theta, \cos\theta, 0) = -\cos\varphi\cos\theta\sin\theta + \cos\varphi\sin\theta\cos\theta = 0.$$

43. From the formulas in (10) in §1.7, we have that

$$x_1 = \rho\sin\varphi_1\sin\varphi_2\cdots\sin\varphi_{n-2}\cos\varphi_{n-1}$$

and

$$x_2 = \rho\sin\varphi_1\sin\varphi_2\cdots\sin\varphi_{n-2}\sin\varphi_{n-1}.$$

Thus when we take the ratio x_2/x_1, everything cancels to leave us with

$$\frac{x_2}{x_1} = \frac{\sin\varphi_{n-1}}{\cos\varphi_{n-1}} = \tan\varphi_{n-1}.$$

47. By part (a) of the previous exercise with $k = n-1$, we have

$$x_1^2 + \cdots + x_{n-1}^2 = \rho^2\sin^2\varphi_1\cdots\sin^2\varphi_{n-(n-1)} = \rho^2\sin^2\varphi_1.$$

Hence

$$(x_1^2 + \cdots + x_{n-1}^2) + x_n^2 = \rho^2\sin^2\varphi_1 + \rho^2\cos^2\varphi_1 = \rho^2.$$

True/False Exercises for Chapter 1

1. False. (The corresponding components must be equal.)
3. False. ($(-4,-3,-3)$ is the displacement vector from P_2 to P_1.)
5. False. (Velocity is a vector, but speed is a scalar.)
7. False. (The particle will be at $(2,-1) + 2(1,3) = (4,5)$.)
9. False. (From the parametric equations, we may read a vector parallel to the line to be $(-2,4,0)$. This vector is not parallel to $(-2,4,7)$.)
11. False. (The line has symmetric form $\frac{x-2}{-3} = y-1 = \frac{z+3}{2}$.)
13. False. (The parametric equations describe a *semicircle* because of the restriction on t.)
15. False. ($\|k\mathbf{a}\| = |k|\,\|\mathbf{a}\|$.)
17. False. (Let $\mathbf{a} = \mathbf{b} = \mathbf{i}$, and $\mathbf{c} = \mathbf{j}$.)
19. True.
21. True. (Check that each point satisfies the equation.)
23. False. (The product *BA* is not defined.)
25. False. ($\det(2A) = 2^n \det A$.)
27. False. (The surface with equation $\rho = 4\cos\varphi$ is a sphere.)
29. True.

Miscellaneous Exercises for Chapter 1

1. *Solution 1.* We add the vectors head-to-tail by parallel translating $\overrightarrow{OP}_2$ so its tail is at the vertex P_1, translating $\overrightarrow{OP}_3$ so that its tail is at the head of the translated $\overrightarrow{OP}_2$, etc. Since each vector $\overrightarrow{OP}_i$ has the same length and, for $i = 2, \ldots, n$, the vector $\overrightarrow{OP}_i$ is rotated $2\pi/n$ from $\overrightarrow{OP}_{i-1}$, the translated vectors will form a closed (regular) n-gon, as the figure below in the case $n = 5$ demonstrates.

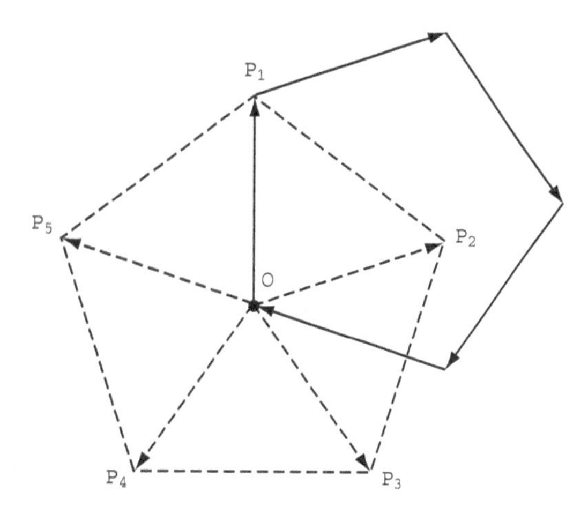

Thus, using head-to-tail addition with the closed n-gon, we see that $\sum_{i=1}^{n} \overrightarrow{OP}_i = \mathbf{0}$.

Solution 2. Suppose that $\sum_{i=1}^{n} \overrightarrow{OP}_i = \mathbf{a} \neq \mathbf{0}$. Imagine rotating the entire configuration through an angle of $2\pi/n$ about the center O of the polygon. The vector $\mathbf{a}$ will have rotated to a different nonzero vector $\mathbf{b}$. However, the original polygon will have rotated to an identical polygon (except for the vertex labels), so the new vector sum $\sum_{i=1}^{n} \overrightarrow{OP}_i$ must be unchanged. Hence $\mathbf{a} = \mathbf{b}$, which is a contradiction. Thus $\mathbf{a} = \mathbf{0}$.

5. **(a)** The desired line must pass through the midpoint of $\overline{P_1P_2}$, which has coordinates $\left(\frac{-1+5}{2}, \frac{3-7}{2}\right) = (2, -2)$. The line must also be perpendicular to $\overrightarrow{P_1P_2}$. The vector $\overrightarrow{P_1P_2}$ is $(5+1, -7-3) = (6, -10)$. A vector perpendicular to this must satisfy $(6, -10) \cdot (a_1, a_2) = 0$ so $3a_1 - 5a_2 = 0$ Hence $\mathbf{a} = (5, 3)$ will serve. A vector parametric equation for the line is $\mathbf{l}(t) = (2, -2) + t(5, 3)$, yielding

$$\begin{cases} x = 5t + 2 \\ y = 3t - 2. \end{cases}$$

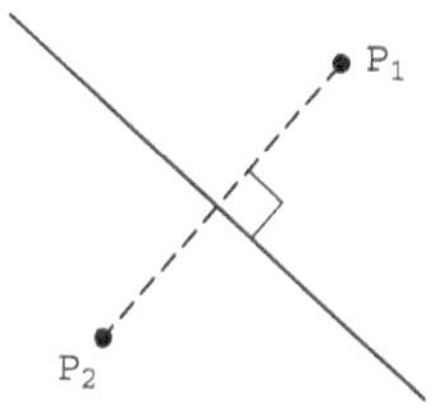

(b) We generalize part (a). Midpoint of $\overline{P_1P_2}$ is $\left(\frac{a_1+b_1}{2}, \frac{a_2+b_2}{2}\right)$. Vector $\overrightarrow{P_1P_2}$ is $(b_1 - a_1, b_2 - a_2)$. A vector $\mathbf{v}$ perpendicular to $\overrightarrow{P_1P_2}$ satisfies $(b_1 - a_1, b_2 - a_2) \cdot \mathbf{v} = 0$ We may therefore take $\mathbf{v}$ to be $\mathbf{v} = (b_2 - a_2, a_1 - b_1)$ so $\mathbf{l}(t) = \left(\frac{a_1+b_1}{2}, \frac{a_2+b_2}{2}\right) + t(b_2 - a_2, a_1 - b_1)$ yielding

$$x = (b_2 - a_2)t + \frac{a_1 + b_1}{2}$$

$$y = (a_1 - b_1)t + \frac{a_2 + b_2}{2}.$$

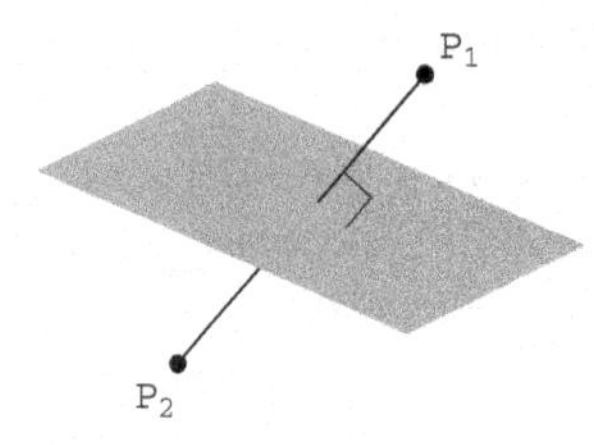

9. **(a)** No. $\mathbf{a} \cdot \mathbf{b} = \mathbf{a} \cdot \mathbf{c}$ just means that the angle between vectors **a** and **b** and the angle between vectors **a** and **b** have the same cosine. If you would prefer, rewrite the equation as $\mathbf{a} \cdot (\mathbf{b} - \mathbf{c}) = 0$ and you can see that what this says is that one of the following is true: vector **a** is orthogonal to the vector $\mathbf{b} - \mathbf{c}$ or $\mathbf{a} = \mathbf{0}$ or $\mathbf{b} - \mathbf{c} = \mathbf{0}$.

(b) No. Use the distributive property of cross products to rewrite the equation as $\mathbf{a} \times (\mathbf{b} - \mathbf{c}) = \mathbf{0}$. This could be true if **a** is parallel to $\mathbf{b} - \mathbf{c}$ or if $\mathbf{a} = \mathbf{0}$ or if $\mathbf{b} - \mathbf{c} = \mathbf{0}$.

13. First note that vectors normal to the respective planes are given by:

$$\begin{aligned}
\mathbf{a} &= 2\mathbf{i} + 3\mathbf{j} - \mathbf{k} && \text{for plane (a)},\\
\mathbf{b} &= -6\mathbf{i} + 4\mathbf{j} - 2\mathbf{k} && \text{for plane (b)},\\
\mathbf{c} &= \mathbf{i} + \mathbf{j} - \mathbf{k} && \text{for plane (c)},\\
\mathbf{d} &= 10\mathbf{i} + 15\mathbf{j} - 5\mathbf{k} && \text{for plane (d)},\\
\mathbf{e} &= 3\mathbf{i} - 2\mathbf{j} + \mathbf{k} && \text{for plane (e)}.
\end{aligned}$$

It is easy to see that $\mathbf{d} = 5\mathbf{a}$ and $\mathbf{b} = -2\mathbf{e}$ and that **c** is not a scalar multiple of any of the other vectors (also that **b** and **d** are not multiples of one another). Hence planes (a) and (d) must be at least parallel; so must planes (b) and (e). In the case of (b) and (e) note that the equation for (b) may be written as

$$-2(3x - 2y + z) = -2(1).$$

That is, the equation for (b) may be transformed into that for (e) by dividing terms by -2. Hence (b) and (e) are equations for the same plane. In the case of (a) and (d), note that $(0, 0, -3)$ lies on plane (a), but not on (d). Hence (a) and (d) are parallel, but not identical. Finally, it is easy to check that $\mathbf{c} \cdot \mathbf{e} = 3 - 2 - 1 = 0$. Thus the normal vectors to planes (c) and (e) are perpendicular, so that the corresponding planes are perpendicular as well. ($\mathbf{c} \cdot \mathbf{a} = 2 + 3 + 1 = 6 \neq 0$, so plane (c) is perpendicular to neither plane (a) nor (d).)

17. We are assuming that the plane Π contains the vectors **a, b, c,** and **d**. The vector $\mathbf{n}_1 = \mathbf{a} \times \mathbf{b}$ is orthogonal to Π, and the vector $\mathbf{n}_2 = \mathbf{c} \times \mathbf{d}$ is orthogonal to Π. So the vectors $\mathbf{n}_1$ and $\mathbf{n}_2$ are parallel. This means that $\mathbf{n}_1 \times \mathbf{n}_2 = \mathbf{0}$.

19. **(a)** The vertices are given so that if they are connected in order *ABDC* we will sketch a parallelogram. From Exercise 18 we could say that

$$\begin{aligned}
\text{Area} &= \sqrt{\|\overrightarrow{AB}\|^2\,\|\overrightarrow{AC}\|^2 - (\overrightarrow{AB} \cdot \overrightarrow{AC})^2}\\
&= [\|(4-1, -1-3, 3+1)\|^2\,\|(2-1, 5-3, 2+1)\|^2\\
&\quad - ((4-1, -1-3, 3+1) \cdot (2-1, 5-3, 2+1))^2]^{1/2}\\
&= \sqrt{\|(3, -4, 4)\|^2\,\|(1, 2, 3)\|^2 - ((3, -4, 4) \cdot (1, 2, 3))^2}\\
&= \sqrt{(41)(14) - (7^2)} = \sqrt{525} = 5\sqrt{21}.
\end{aligned}$$

(b) When we project the parallelogram in the xy-plane we get the same points with the z coordinate equal to 0. We do the same calculation as in part a with the new vectors:

$$\begin{aligned}
\text{Area} &= \sqrt{\|\overrightarrow{AB}\|^2\,\|\overrightarrow{AC}\|^2 - (\overrightarrow{AB} \cdot \overrightarrow{AC})^2}\\
&= \sqrt{\|(4-1, -1-3, 0)\|^2\,\|(2-1, 5-3, 0)\|^2 - ((4-1, -1-3, 0) \cdot (2-1, 5-3, 0))^2}\\
&= \sqrt{\|(3, -4, 0)\|^2\,\|(1, 2, 0)\|^2 - ((3, -4, 0) \cdot (1, 2, 0))^2}\\
&= \sqrt{(25)(5) - (5^2)} = \sqrt{100} = 10.
\end{aligned}$$

23. **(a)** The vector $\overrightarrow{AB} \times \overrightarrow{AC} = \mathbf{0}$ if and only if $\overrightarrow{AB}$ is parallel to $\overrightarrow{AC}$. This happens if and only if A, B, and C are collinear.

(b) We note that $\overrightarrow{CD} \neq \mathbf{0}$ since C and D are distinct points. Then $(\overrightarrow{AB} \times \overrightarrow{AC}) \cdot \overrightarrow{CD} = 0$ if and only if $\overrightarrow{AB} \times \overrightarrow{AC} = \mathbf{0}$ or $\overrightarrow{AB} \times \overrightarrow{AC}$ is perpendicular to $\overrightarrow{CD}$. The first case occurs exactly when A, B, and C are collinear (so A, B, C and D are coplanar). In the second case, $\overrightarrow{AB} \times \overrightarrow{AC}$ is perpendicular to the plane containing A, B, and C and so $\overrightarrow{CD}$ can only be perpendicular to it if and only if D lies in this plane as well.

25. The equation $\mathbf{a} \times \mathbf{x} = \mathbf{b}$ tells us that $\mathbf{x}$ points in the direction of $\mathbf{b} \times \mathbf{a}$. Now we have to determine the length of $\mathbf{x}$. We can choose any vector in the direction of $\mathbf{x}$. For convenience, let $\mathbf{y}$ be the unit vector in direction of $\mathbf{x}$:

$$\mathbf{y} = \frac{\mathbf{b} \times \mathbf{a}}{\|\mathbf{b} \times \mathbf{a}\|}.$$

The angle between $\mathbf{a}$ and $\mathbf{x}$ is the same as that between $\mathbf{a}$ and $\mathbf{y}$ so

$$\frac{\mathbf{a} \cdot \mathbf{y}}{\|\mathbf{a}\|} = \frac{\mathbf{a} \cdot \mathbf{x}}{\|\mathbf{a}\| \, \|\mathbf{x}\|} = \frac{c}{\|\mathbf{a}\| \, \|\mathbf{x}\|}.$$

So if $c \neq 0$,

$$\|\mathbf{x}\| = \frac{c}{\mathbf{a} \cdot \mathbf{y}}, \quad \text{and,} \quad \mathbf{x} = \left(\frac{c}{\mathbf{a} \cdot \mathbf{y}}\right)\mathbf{y}.$$

If $c = 0$ then $\mathbf{a}$ is orthogonal to $\mathbf{x}$ (and $\mathbf{y}$). Use the fact that

$$\|\mathbf{b}\| = \|\mathbf{a} \times \mathbf{x}\| = \|\mathbf{a}\| \, \|\mathbf{x}\| \sin\theta = \|\mathbf{a}\| \, \|\mathbf{x}\| \sin \pi/2 = \|\mathbf{a}\| \, \|\mathbf{x}\|.$$

So when $c = 0$,

$$\|\mathbf{x}\| = \frac{\|\mathbf{b}\|}{\|\mathbf{a}\|} \quad \text{and} \quad \mathbf{x} = \left(\frac{\|\mathbf{b}\|}{\|\mathbf{a}\|}\right)\mathbf{y}.$$

31. **(a)**

$$A = \begin{bmatrix} 1 & 1 \\ 0 & 1 \end{bmatrix}, A^2 = \begin{bmatrix} 1 & 2 \\ 0 & 1 \end{bmatrix}, A^3 = \begin{bmatrix} 1 & 3 \\ 0 & 1 \end{bmatrix}, A^4 = \begin{bmatrix} 1 & 4 \\ 0 & 1 \end{bmatrix}$$

(b) It seems reasonable to guess that

$$A^n = \begin{bmatrix} 1 & n \\ 0 & 1 \end{bmatrix}.$$

(c) We need only show the inductive step:

$$A^{n+1} = AA^n = \begin{bmatrix} 1 & 1 \\ 0 & 1 \end{bmatrix}\begin{bmatrix} 1 & n \\ 0 & 1 \end{bmatrix} = \begin{bmatrix} 1 & n+1 \\ 0 & 1 \end{bmatrix}.$$

37. Look at the second part of the answers in Exercises 34 and 35. The only difference is that we are changing the distance from the center of the moving wheel to P from b to c. The formula for a hypotrochoid is:

$$(a-b)(\cos t, \sin t) + c\left(\cos\left(\frac{(a-b)t}{b}\right), -\sin\left(\frac{(a-b)t}{b}\right)\right).$$

In parametric form, the formulas for a hypotrochoid are:

$$x = (a-b)\cos t + c\cos\left(\frac{(a-b)t}{b}\right), \; y = (a-b)\sin t - c\sin\left(\frac{(a-b)t}{b}\right).$$

The formula for an epitrochoid is:

$$(a+b)(\cos t, \sin t) - c\left(\cos\left(\frac{(a+b)t}{b}\right), \sin\left(\frac{(a+b)t}{b}\right)\right).$$

In parametric form, the formulas for an epitrochoid are:

$$x = (a+b)\cos t - c\cos\left(\frac{(a+b)t}{b}\right), \; y = (a+b)\sin t - c\sin\left(\frac{(a+b)t}{b}\right).$$

39. Parts (a), (b), and (d) are pictured below top, left, and right. They look very similar to the graphs from the previous exercise.

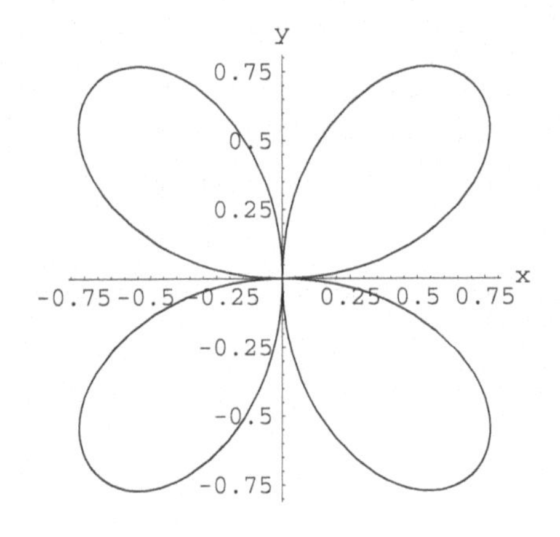

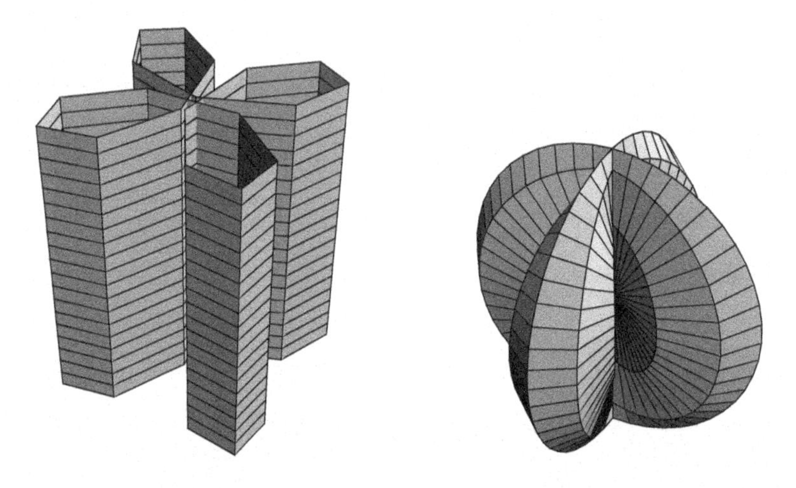

(c) This looks very different from its counterpart for Exercise 38. It looks like a dented sphere.

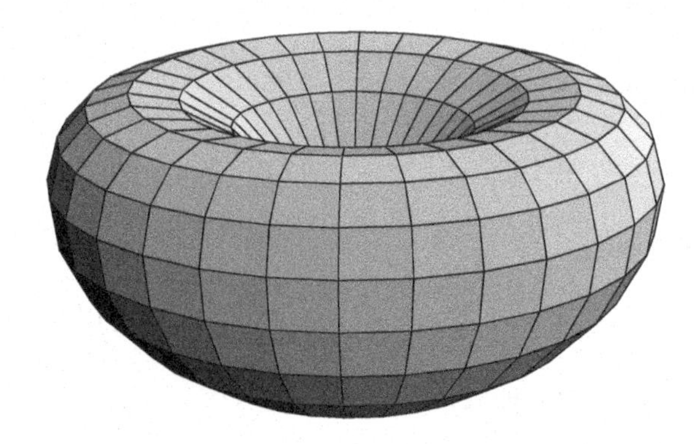

43. **(a)** The curve is a spiral and is pictured below left.

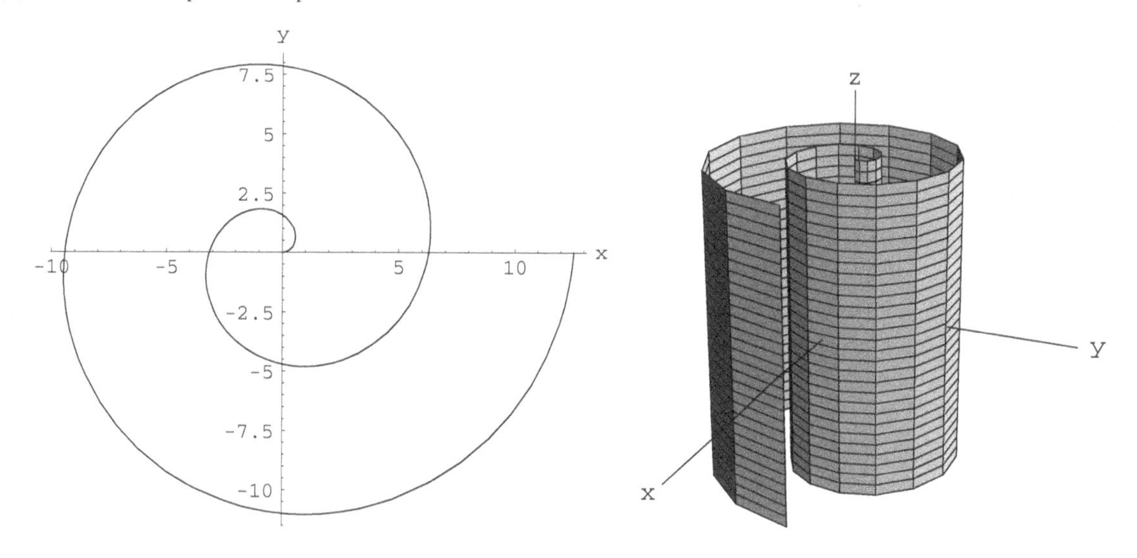

(b) The cylinder based on the spiral in part (a) is shown above right.

(c) Because only part of the spiral is used, the resulting surface is a dimpled ball.

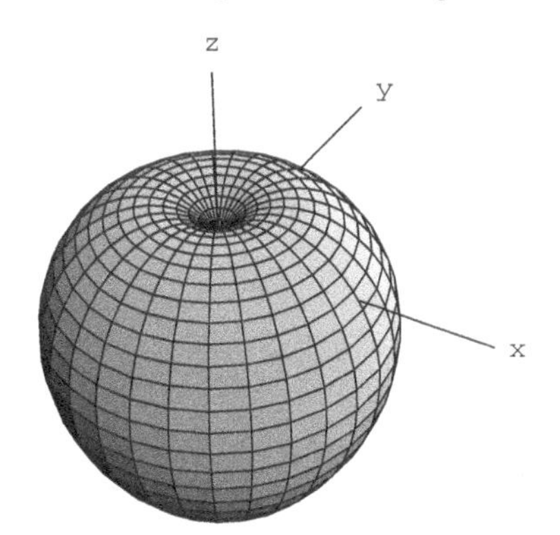

(d) Finally, we see a lovely and intricate shell-like surface.

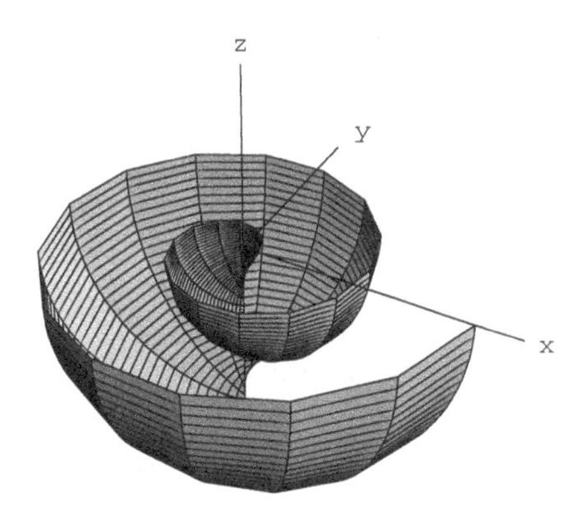

Chapter 2

Differentiation in Several Variables

2.1 Functions of Several Variables; Graphing Surfaces

1. $f\colon \mathbf{R} \to \mathbf{R}\colon x \mapsto 2x^2 + 1$
(a) Domain $f = \{x \in \mathbf{R}\}$, Range $f = \{y \in \mathbf{R} | y \geq 1\}$.
(b) No. For instance $f(1) = 3 = f(-1)$.
(c) No. For instance if $y = 0$, there is no x such that $f(x) = 0$.

7. Domain $\mathbf{f} = \{(x, y) \in \mathbf{R}^2 | y \neq 1\}$, Range $\mathbf{f} = \{(x, y, z) \in \mathbf{R}^3 | y \neq 0, y^2 z = (xy - y - 1)^2 + (y + 1)^2\}$.

11. **(a)** $\mathbf{f}(\mathbf{x}) = -2\mathbf{x}/\|\mathbf{x}\|$.
(b) The component functions are

$$f_1(x, y, z) = \frac{-2x}{\sqrt{x^2 + y^2 + z^2}}, \; f_2(x, y, z) = \frac{-2y}{\sqrt{x^2 + y^2 + z^2}}, \quad \text{and} \quad f_3(x, y, z) = \frac{-2z}{\sqrt{x^2 + y^2 + z^2}}.$$

15. For $c > 0$ the level sets are circles centered at the origin of radius $\sqrt{c}$. For $c = 0$ the level set is just the origin. There are no values corresponding to $c < 0$. Note that the curves get closer together, indicating that we are climbing faster as we head out radially from the origin. The second figure below shows the plot of the level curves shaded to indicate the height of the level set (lighter is higher). The surface is therefore a paraboloid symmetric about the z-axis. We show it with and without the surface filled in.

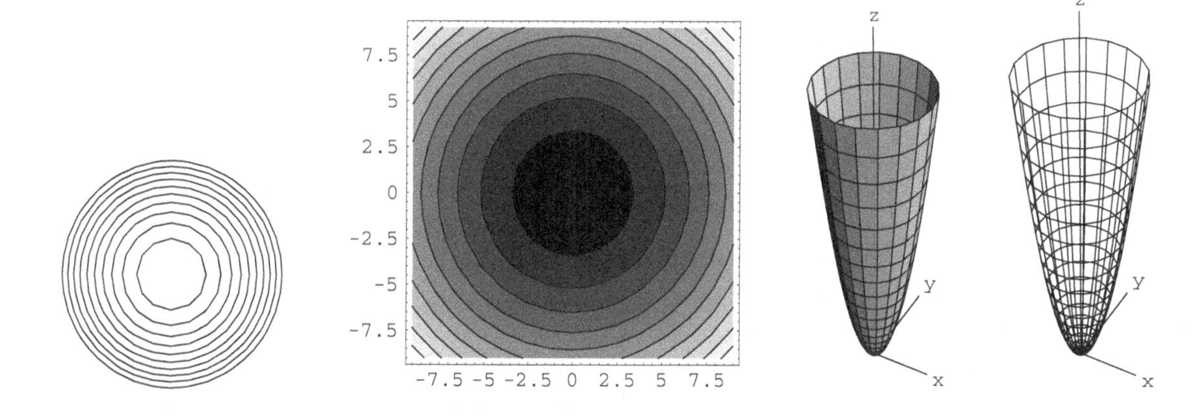

19. The graphs $xy = c$ are hyperbolas (unless $c = 0$ in which case it is the union of the two axes). When x and y are both positive the height of the level curves are positive and so the hyperboloid is increasing as we head away from the origin radially in either the first or third quadrant. When x and y are of different signs, the heights of the level curves are negative and so the hyperboloid is decreasing as we head out radially in either the second or fourth quadrant.

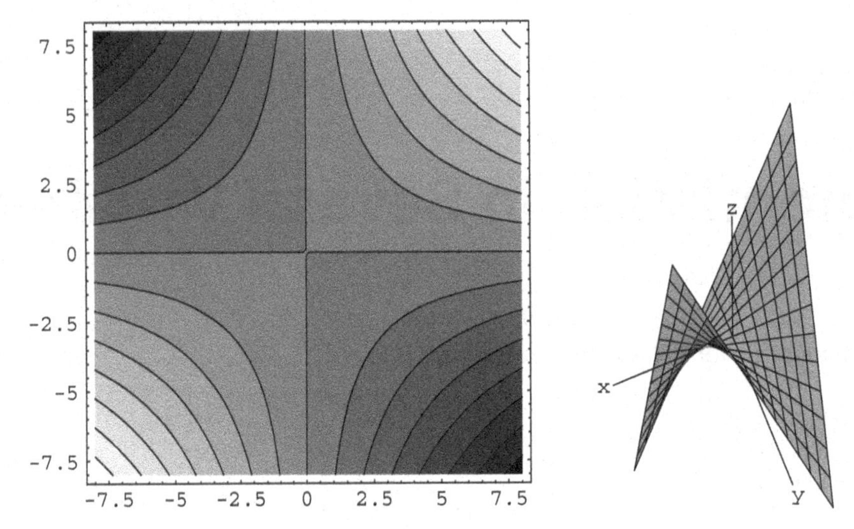

21. We have a problem when $y = 0$. When $k < 0$, the section by $x = k$ looks like the hyperbola in the figure on the left, when $k > 0$, the section looks like the hyperbola in the figure on the right:

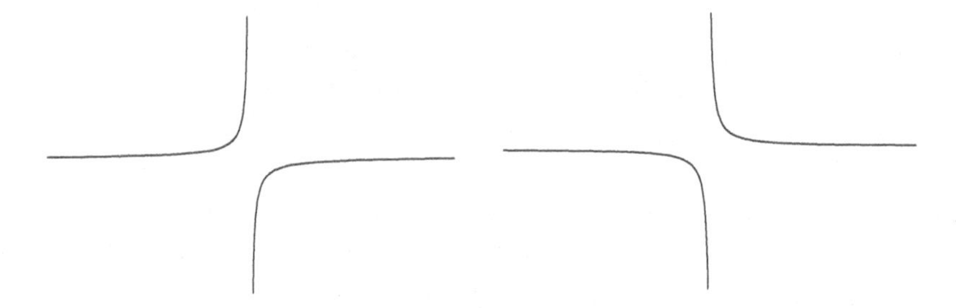

You can see that as $y \to 0$ from either side, along a line where x is constant and not 0, the z values won't match up. We are going to get a tear down the line $y = 0$. The level sets look like:

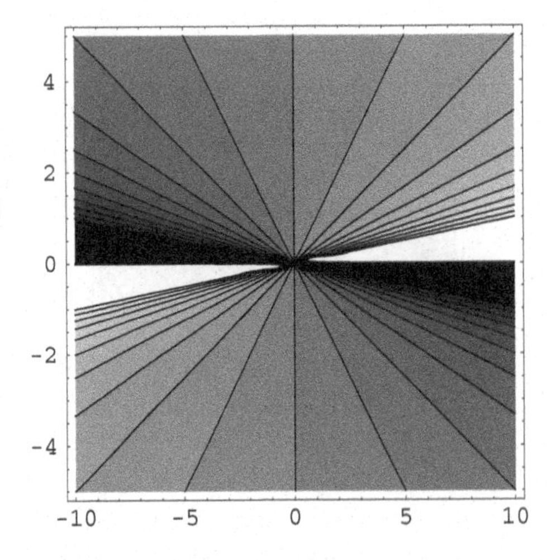

Notice that you can see that tear on the right center part of the above graph. The solid black and solid white areas which are on either side of the x-axis point to the behavior around the tear.

Graph each side of the x-axis and you will see the following piece of the surface:

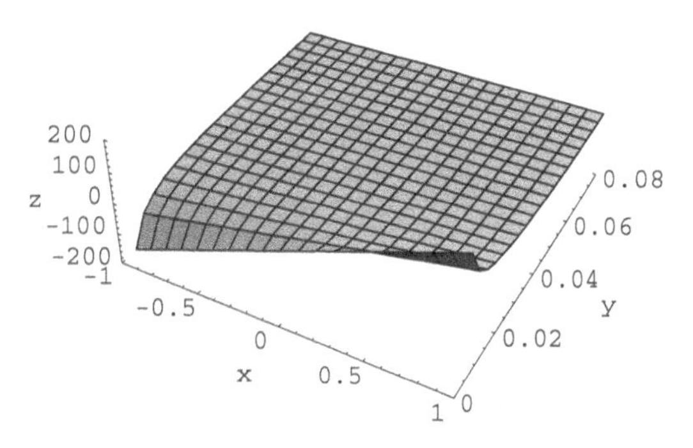

Our final surface is what you get when you try to glue two of those together:

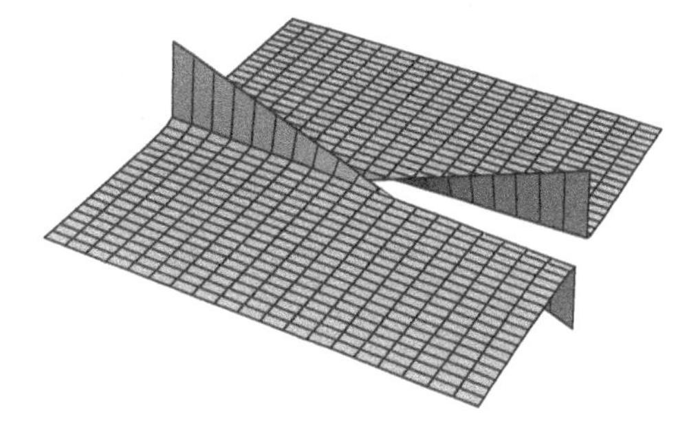

25. Figures below:

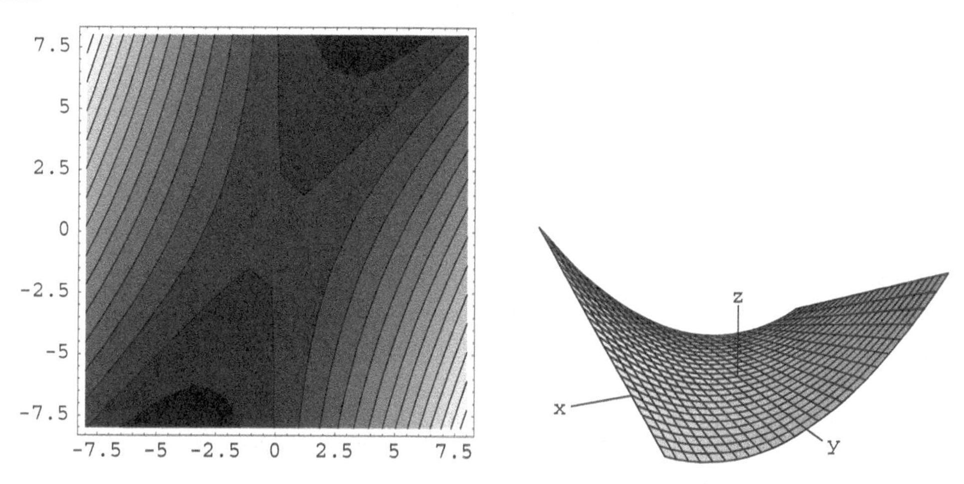

31. They can't intersect—even though they may sometimes appear to. Say that two different level curves $f(x, y) = c_1$ and $f(x, y) = c_2$ where $c_1 \neq c_2$ intersect at some point (a, b). Then $f(a, b)$ would have assigned to it two non-equal values. This can't happen for a function (it's our vertical line test). On the other hand, if the limit as you approach (a, b) along different paths is different, those level curves may appear to intersect at (a, b) no matter how good the resolution on your contour plot.

35. The level surfaces at level $w = c$ are nested ellipsoids.

39. The ellipsoid is pictured below left. To see why you couldn't express the surface as one function $z = f(x, y)$, look for example at the intersection of the ellipsoid and the plane $y = 0$ pictured below on the right.

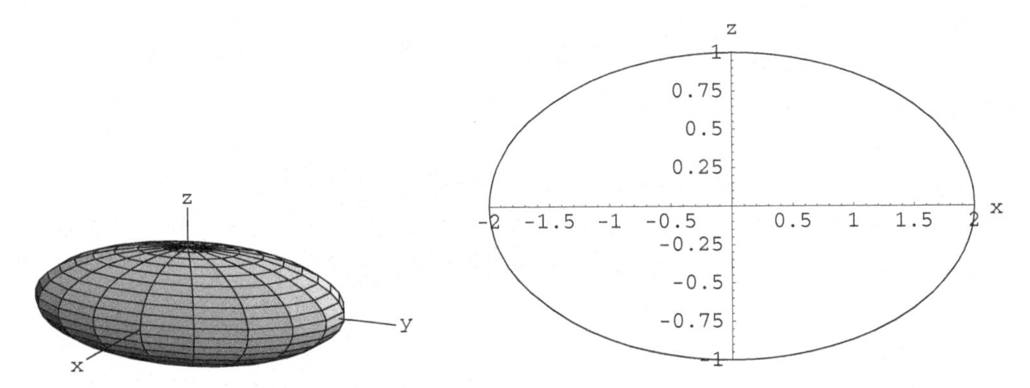

You can see that for $-2 < x < 2$ there correspond two values of z. We could express the top portion of the ellipsoid as $f(x, y) = \sqrt{1 - (x^2/4 + y^2/9)}$ and the bottom portion as $g(x, y) = -\sqrt{1 - (x^2/4 + y^2/9)}$.

41. Here we get a cone with axis of symmetry the x-axis. The figure is shown below.

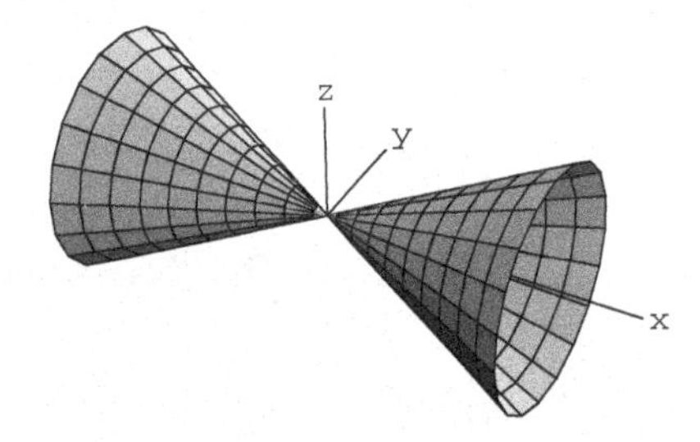

47. This is the equation of an elliptic cone with vertex at $(1, -1, -3)$. The graph is shown below.

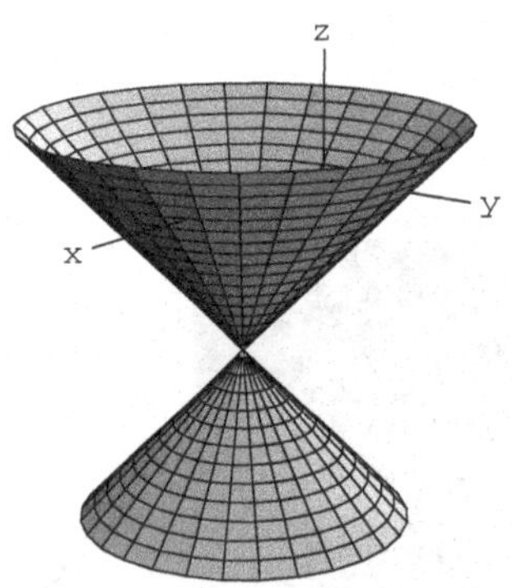

51. The equation is equivalent to $z - 1 = (x - 3)^2 + 2y^2$, and so is an elliptic paraboloid.

2.2 Limits

1. This is an annulus which doesn't include its inner or outer boundary and so is **open**.

5. This may be a bit harder to see. This is the union of an infinite open strip in the plane $(-1 < x < 1)$ and a closed line in the plane $(x = 2)$ and so is **neither open nor closed**.

9. The limit does not exist.

$$\lim_{(x,y)\to(0,0)} \frac{(x+y)^2}{x^2+y^2} = \lim_{(x,y)\to(0,0)} \frac{x^2+2xy+y^2}{x^2+y^2} = 1 + \lim_{(x,y)\to(0,0)} \frac{2xy}{x^2+y^2}.$$

When $x = y$,

$$1 + \lim_{(x,y)\to(0,0)} \frac{2xy}{x^2+y^2} = 1 + \lim_{x\to 0} \frac{2x^2}{x^2+x^2} = 1 + 1 = 2.$$

When $x = 0$,

$$1 + \lim_{(0,y)\to(0,0)} \frac{2xy}{x^2+y^2} = 1 + \lim_{y\to 0} \frac{0}{y^2} = 1.$$

13. Just as with limits in first semester Calculus, this is begging to be simplified.

$$\lim_{(x,y)\to(0,0)} \frac{x^2+2xy+y^2}{x+y} = \lim_{(x,y)\to(0,0)} \frac{(x+y)^2}{x+y} = \lim_{(x,y)\to(0,0)} (x+y) = 0.$$

17. This is another standard trick from first year Calculus.

$$\lim_{(x,y)\to(0,0),x\neq y} \frac{x^2-xy}{\sqrt{x}-\sqrt{y}} = \lim_{(x,y)\to(0,0),x\neq y} \frac{x(x-y)}{\sqrt{x}-\sqrt{y}} = \lim_{(x,y)\to(0,0),x\neq y} \frac{x(\sqrt{x}+\sqrt{y})(\sqrt{x}-\sqrt{y})}{\sqrt{x}-\sqrt{y}}$$
$$= \lim_{(x,y)\to(0,0),x\neq y} x(\sqrt{x}+\sqrt{y}) = 0.$$

23. Our goal is to evaluate $\lim_{(x,y)\to(0,0)} \dfrac{x^4y^4}{(x^2+y^4)^3}$ or explain why the limit fails to exist. We divide the answer into parts to make it easier to follow—there are no corresponding parts (a)–(d) in the text.

(a) If you evaluate the limit along the lines $x = 0$ and $y = 0$ the limit is 0. We might be tempted to guess that $\lim_{(x,y)\to(0,0)} f(x,y) = 0$ but as we saw in Exercise 14, we could get a limit of 0 along the paths $x = 0$ and $y = 0$ but perhaps not along $x = y$.

(b) So now let's follow the line $y = mx$ into the origin and see where f heads off to.

$$\begin{aligned}
\lim_{(x,y)\to(0,0),y=mx} \frac{x^4y^4}{(x^2+y^4)^3} &= \lim_{x\to 0} \frac{x^4(mx)^4}{(x^2+(mx)^4)^3} \\
&= \lim_{x\to 0} \frac{m^4x^8}{(x^2(1+m^4x^2))^3} \\
&= m^4 \lim_{x\to 0} \frac{x^8}{(x^6)(1+m^4x^2)^3} \\
&= m^4 \lim_{x\to 0} \frac{x^2}{(1+m^4x^2)^3} = 0.
\end{aligned}$$

This means then if we head into the origin along any straight line the limit of f is 0. *Here is the point of this problem*: If we head into the origin in any constant direction, the limit of f is 0 and yet $\lim_{(x,y)\to(0,0)} f(x,y)$ does not exist!

(c) For the limit to exist f must approach the same number no matter what path we choose to take to the origin. So let's approach along the parabola $x = y^2$.

$$\begin{aligned}
\lim_{(x,y)\to(0,0),x=y^2} \frac{x^4y^4}{(x^2+y^4)^3} &= \lim_{y\to 0} \frac{(y^2)^4y^4}{((y^2)^2+y^4)^3} \\
&= \lim_{y\to 0} \frac{y^{12}}{(2y^4)^3} \\
&= \lim_{y\to 0} \frac{y^{12}}{8y^{12}} = \frac{1}{8}.
\end{aligned}$$

(d) So we get different answers for $\lim_{(x,y)\to(0,0)} \dfrac{x^4y^4}{(x^2+y^4)^3}$ depending on what path we follow into the origin. So the limit does not exist.

25. Below see two graphs of the function. You actually get most of the picture from the three-dimensional graph—except that it looks as if things are joined smoothly. The contour plot shows the dramatic problems near the origin. Particularly if you look along the vertical line $x = 0$ you'll see that the limit does not exist at the origin.

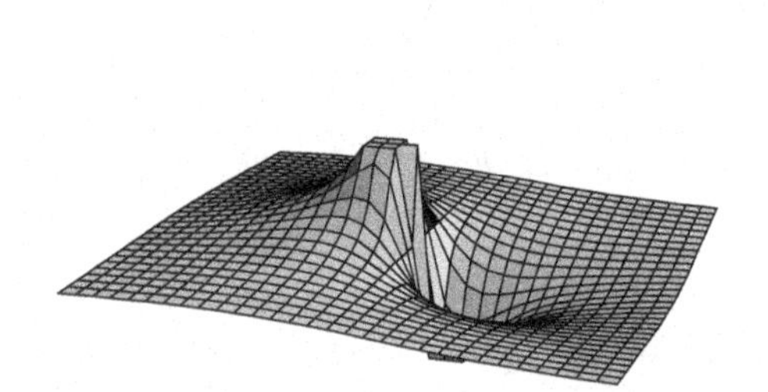

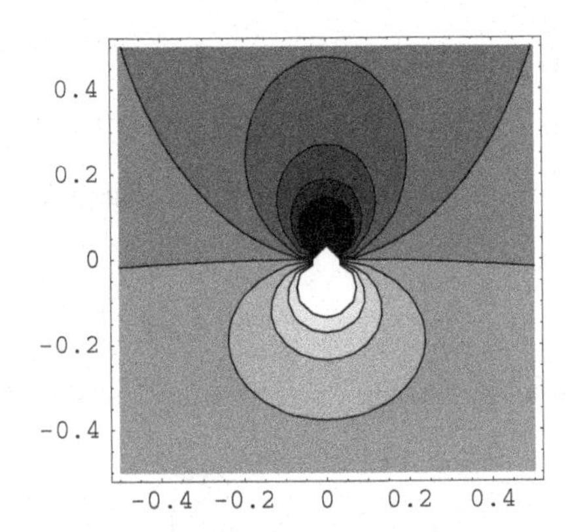

Analytically, look at the path $x = 0$. Here we're looking at the graph of $z = -1/y$. The limits as we approach from positive and negative y values is $\pm\infty$ so no limit exists.

29.

$$\lim_{(x,y)\to(0,0)} \frac{x^2}{x^2+y^2} = \lim_{r\to 0} \frac{r^2\cos^2\theta}{r^2} = \lim_{r\to 0}\cos^2\theta = \cos^2\theta$$

Limit does not exist as the result depends on θ.

31. We have

$$\begin{aligned}
&\lim_{(x,y)\to(0,0)} \frac{x^5+y^4-3x^3y+2x^2+2y^2}{x^2+y^2} \\
&= \lim_{r\to 0} \frac{r^5\cos^5\theta + r^4\sin^4\theta - 3r^4\cos^3\theta\sin\theta + 2r^2\cos^2\theta + 2r^2\sin^2\theta}{r^2\cos^2\theta + r^2\sin^2\theta} \\
&= \lim_{r\to 0} \frac{r^2(r^3\cos^5\theta + r^2\sin^4\theta - 3r^2\cos^3\theta\sin\theta + 2)}{r^2} \\
&= \lim_{r\to 0} \left[r^2(r\cos^5\theta + \sin^4\theta - 3\cos^3\theta\sin\theta) + 2\right]
\end{aligned}$$

Note that $-1 \le \cos^n\theta \le 1$ when n is odd, $-1 \le \sin\theta \le 1$, and $0 \le \sin^m\theta \le 1$ when m is even. Thus we have that

$$r^2(-r+0-3)+2 \le r^2(r\cos^5\theta + \sin^4\theta - 3\cos^3\theta\sin\theta) + 2 \le r^2(r+1+3)+2.$$

Now

$$\lim_{r\to 0}\left[-r^2(r+3)+2\right] = \lim_{r\to 0}\left[r^2(r+4)+2\right] = 2;$$

thus $\lim_{r\to 0}\left[r^2(r\cos^5\theta + \sin^4\theta - 3\cos^3\theta\sin\theta) + 2\right] = 2$ by squeezing.

35.

$$\begin{aligned}
\lim_{(x,y,z)\to(0,0,0)} \frac{xyz}{x^2+y^2+z^2} &= \lim_{\rho\to 0} \frac{(\rho\sin\varphi\cos\theta)(\rho\sin\varphi\sin\theta)(\rho\cos\varphi)}{\rho^2} \\
&= \lim_{\rho\to 0} \rho\sin^2\varphi\cos\varphi\cos\theta\sin\theta = 0
\end{aligned}$$

39. This is a polynomial and is continuous everywhere.

43. The only place we need to check is the origin. We need to show that the limit of f as we approach $(0, 0)$ is 0. If we add and subtract y^2 to the numerator we find that:

$$\lim_{(x,y)\to(0,0)} \frac{x^2-y^2}{x^2+y^2} = 1 - 2\lim_{(x,y)\to(0,0)} \frac{y^2}{x^2+y^2}.$$

In Exercise 16 we showed that this limit doesn't exist (in this case you get two different answers if you follow the paths $y = 0$ and $y = x$) and so f is not continuous at $(0, 0)$.

47. Here you can view f as being a function $\mathbf{R}^3 \to \mathbf{R}$; then $f(x_1, x_2, x_3) = 2x_1 - 3x_2 + x_3$ which is linear in x_1, x_2, and x_3 and therefore continuous.

51. This is just a generalization of Exercise 50.

(a)

$$\delta > \|(x,y)-(x_0,y_0)\| = \sqrt{(x-x_0)^2+(y-y_0)^2} \geq \sqrt{(x-x_0)^2} = |x-x_0|.$$

And

$$\delta > \|(x,y)-(x_0,y_0)\| = \sqrt{(x-x_0)^2+(y-y_0)^2} \geq \sqrt{(y-y_0)^2} = |y-y_0|.$$

(b) Assume that $\|(x,y)-(x_0,y_0)\| < \delta$, then follow the steps in part (b) of Exercise 50:

$$\begin{aligned} |f(x,y)-(Ax_0+By_0+C)| &= |Ax+By+C-(Ax_0+By_0+C)| \\ &= |A(x-x_0)+B(y-y_0)| \leq |A(x-x_0)|+|B(y-y_0)| \\ &= |A||x-x_0|+|B||y-y_0| < |A|\delta+|B|\delta = (|A|+|B|)\delta. \end{aligned}$$

(c) Now we're ready to put this together: For any $\varepsilon > 0$, if $0 < \|(x,y)-(x_0,y_0)\| < \varepsilon/(|A|+|B|)$, then $|f(x,y)-(Ax_0+By_0+C)| < \varepsilon$. In other words,

$$\lim_{(x,y)\to(x_0,y_0)} f(x,y) = Ax_0+By_0+C.$$

2.3 The Derivative

1. $f(x,y)=xy^2+x^2y$, so $\partial f/\partial x = y^2+2xy$, and $\partial f/\partial y = 2xy+x^2$.

7. $f(x,y)=\cos x^3y$, so $\dfrac{\partial f}{\partial x} = (-\sin x^3y)(3yx^2) = -3x^2y\sin x^3y$ and $\dfrac{\partial f}{\partial y} = -x^3\sin x^3y$.

11. $F(x,y,z)=\dfrac{x-y}{y+z}$, so $\dfrac{\partial F}{\partial x}=\dfrac{1}{y+z}$,

$$\frac{\partial F}{\partial y} = \frac{(y+z)(-1)-(x-y)(1)}{(y+z)^2} = -\frac{x+z}{(y+z)^2},$$

and

$$\frac{\partial F}{\partial z} = \frac{(y+z)(0)-(x-y)(1)}{(y+z)^2} = \frac{y-x}{(y+z)^2}.$$

15. $F(x,y,z)=\dfrac{x+y+z}{(1+x^2+y^2+z^2)^{3/2}}$

$$\begin{aligned} F_x(x,y,z) &= \frac{(1+x^2+y^2+z^2)^{3/2}-(x+y+z)(3/2)(1+x^2+y^2+z^2)^{1/2}(2x)}{(1+x^2+y^2+z^2)^3} \\ &= \frac{1-2x^2+y^2+z^2-3xy-3xz}{(1+x^2+y^2+z^2)^{5/2}} \\ F_y(x,y,z) &= \frac{1+x^2-2y^2+z^2-3xy-3yz}{(1+x^2+y^2+z^2)^{5/2}}, \text{ and} \\ F_z(x,y,z) &= \frac{1+x^2+y^2-2z^2-3xz-3yz}{(1+x^2+y^2+z^2)^{5/2}}. \end{aligned}$$

19. $f(x,y)=\dfrac{x-y}{x^2+y^2+1}$, so

$$\begin{aligned} \nabla f(x,y) &= \left(\frac{(x^2+y^2+1)(1)-(x-y)(2x)}{(x^2+y^2+1)^2}, \frac{(x^2+y^2+1)(-1)-(x-y)(2y)}{(x^2+y^2+1)^2}\right) \\ &= \left(\frac{-x^2+y^2+1+2xy}{(x^2+y^2+1)^2}, \frac{-x^2+y^2-1-2xy}{(x^2+y^2+1)^2}\right). \end{aligned}$$

So

$$\nabla f(2,-1) = \left(-\frac{6}{36}, \frac{0}{36}\right) = \left(-\frac{1}{6}, 0\right).$$

21. $f(x, y, z) = xy + y\cos z - x\sin yz$, so $\nabla f(x, y, z) = (y - \sin yz, x + \cos z - xz\cos yz, -y\sin z - xy\cos yz)$. So,

$$\begin{aligned}\nabla f(2, -1, \pi) &= (-1 - \sin(-\pi), 2 + \cos(\pi) - 2(\pi)\cos(-\pi), \sin(\pi) + 2\cos(-\pi)) \\ &= (-1, 1 + 2\pi, -2).\end{aligned}$$

25. $f(x, y, z) = \dfrac{xy^2 - x^2 z}{y^2 + z^2 + 1}$, so we have $\dfrac{\partial f}{\partial x} = \dfrac{y^2 - 2xz}{y^2 + z^2 + 1}$ and the quotient rule applied appropriately gives

$$\frac{\partial f}{\partial x} = \frac{(y^2 + z^2 + 1)(2xy) - (xy^2 - x^2 z)(2y)}{(y^2 + z^2 + 1)^2} = \frac{2xy(xz + z^2 + 1)}{(y^2 + z^2 + 1)^2}$$

and

$$\frac{\partial f}{\partial x} = \frac{(y^2 + z^2 + 1)(-x^2) - (xy^2 - x^2 z)(2z)}{(y^2 + z^2 + 1)^2} = \frac{x(xz^2 - xy^2 - 2y^2 z - x)}{(y^2 + z^2 + 1)^2}.$$

Therefore, $\nabla f(-1, 2, 1) = (1, -1/9, 1/9)$.

31. $\mathbf{f}(x, y, z, w) = (3x - 7y + z, 5x + 2z - 8w, y - 17z + 3w)$ so

$$D\mathbf{f}(x, y, z, w) = \begin{bmatrix} 3 & -7 & 1 & 0 \\ 5 & 0 & 2 & -8 \\ 0 & 1 & -17 & 3 \end{bmatrix}.$$

Since all of the entries are constant, the matrix doesn't depend on **a**.

35. $f(x, y, z) = \dfrac{x + y + z}{x^2 + y^2 + z^2}$ is differentiable because the three partials

$$f_x(x, y, z) = \frac{-x^2 + y^2 + z^2 - 2xy - 2xz}{(x^2 + y^2 + z^2)^2}$$

$$f_y(x, y, z) = \frac{x^2 - y^2 + z^2 - 2xy - 2yz}{(x^2 + y^2 + z^2)^2}$$

$$f_z(x, y, z) = \frac{x^2 + y^2 - z^2 - 2xz - 2yz}{(x^2 + y^2 + z^2)^2}$$

are all continuous.

39. Using Theorem 3.3, the equation for the tangent plane is: $z = f(0, 1) + f_x(0, 1)(x) + f_y(0, 1)(y - 1)$. Here $z = e^{x+y}\cos xy$, so $f_x(x, y) = e^{x+y}(\cos xy - y\sin xy)$ and $f_y(x, y) = e^{x+y}(\cos xy - x\sin xy)$. Plugging in we get $z = e + ex + e(y - 1)$ or $z = ex + ey$.

43. Here $f(x, y) = e^{x+y}$ so the partials are $f_x(x, y) = e^{x+y} = f_y(x, y)$.

(a) $h(.1, -.1) = f(0, 0) + (e^0, e^0) \cdot (.1, -.1) = 1$.

(b) $f(.1, -.1) = e^0 = 1$. So the approximation is exact.

47. **(a)** For $(x, y) \neq (0, 0)$ we can find a neighborhood that misses the origin. In this neighborhood

$$f(x, y) = \frac{xy^2 - x^2 y + 3x^3 - y^3}{x^2 + y^2} = x - y + \frac{2x^3}{x^2 + y^2}.$$

We can then easily compute the partials as

$$f_x(x, y) = 1 + \frac{2x^4 + 6x^2 y^2}{(x^2 + y^2)^2} \quad \text{and} \quad f_y(x, y) = -1 - \frac{4x^3 y}{(x^2 + y^2)^2}.$$

(b) Using Definition 3.2 of the partial derivative, if

$$f(x, y) = \begin{cases} x - y + \dfrac{2x^3}{x^2 + y^2} & \text{if } (x, y) \neq (0, 0) \\ 0 & \text{if } (x, y) = (0, 0) \end{cases},$$

then

$$\frac{\partial f}{\partial x}(0, 0) = \lim_{h \to 0} \frac{f(h, 0) - f(0, 0)}{h} = \lim_{h \to 0} \frac{3h}{h} = 3,$$

and

$$\frac{\partial f}{\partial y}(0, 0) = \lim_{h \to 0} \frac{f(0, h) - f(0, 0)}{h} = \lim_{h \to 0} \frac{-h}{h} = -1.$$

49. For the tangent line to $F(x) = x + \sin x$ at $a = \pi/4$ $F'(x) = 1 + \cos x$ so $F'(\pi/4) = 1 + \sqrt{2}/2$. The tangent line is $y = (1 + \sqrt{2}/2)x + (\pi/4 + \sqrt{2}/2 - (1 + \sqrt{2}/2)\pi/4)$. The graph of F and the tangent line near $x = \pi/4$ is shown below.

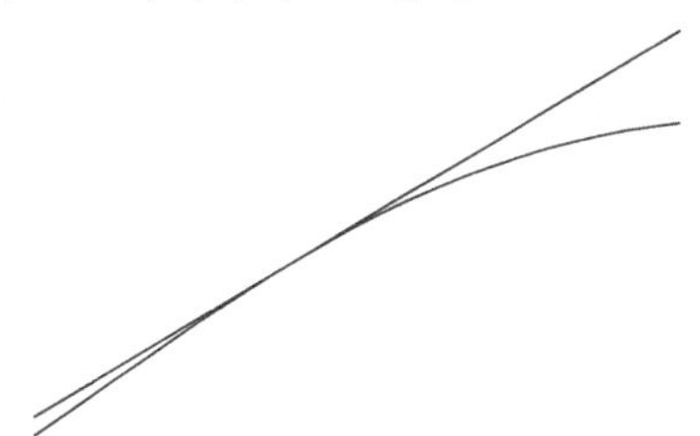

55. **(a)** For the function $f(x, y) = \dfrac{xy}{x^2 + y^2 + 1}$, $f_x(x, y) = \dfrac{-x^2y + y^3 + y}{(x^2 + y^2 + 1)^2}$ and $f_y(x, y) = \dfrac{x^3 - xy^2 + y}{(x^2 + y^2 + 1)^2}$. So at the point $(0, 0)$ these become $f(0, 0) = 0$, $f_x(0, 0) = 0$, and $f_y(0, 0) = 0$. The equation of the tangent plane is $z = 0$.

(b) The surface is shown below left. It is shown with the tangent plane below right.

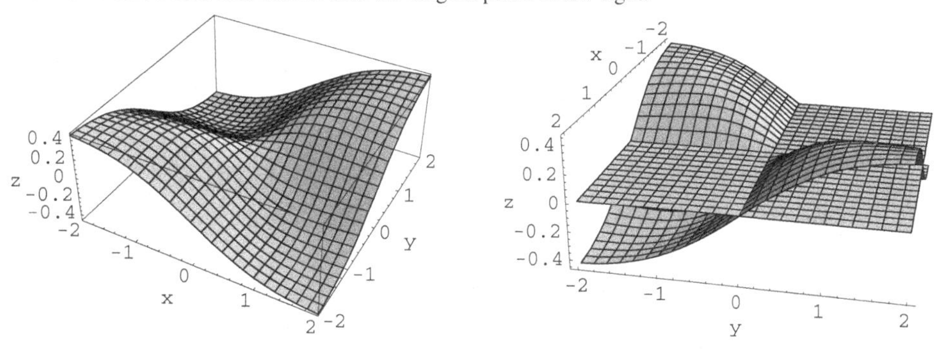

(c) This is the plane that best approximates the surface at that point. But we can see that it's not a very good approximation as you move way in any direction other than the two axes lines. Analytically, the reason is that the partials are continuous in a neighborhood of $(0, 0)$.

59. If $\mathbf{f}(\mathbf{x}) = A\mathbf{x} = \left(\sum_{k=1}^{n} a_{1k}x_k, \sum_{k=1}^{n} a_{2k}x_k, \ldots, \sum_{k=1}^{n} a_{mk}x_k\right)$. Let's look at the entry in row i column j of $D\mathbf{f}(\mathbf{x})$. This will be

$$\frac{\partial f_i}{\partial x_j} = \frac{\partial}{\partial x_j}\left(\sum_{k=1}^{n} a_{ik}x_k\right) = a_{ij}.$$

So $D\mathbf{f}(\mathbf{x}) = A$.

61. **(a)** First, Exercise 60 shows that

$$\lim_{\mathbf{x}\to\mathbf{a}} \frac{\mathbf{f}(\mathbf{x}) - [\mathbf{f}(\mathbf{a}) + A(\mathbf{x} - \mathbf{a})]}{\|\mathbf{x} - \mathbf{a}\|} = \lim_{\mathbf{x}\to\mathbf{a}} \frac{\mathbf{f}(\mathbf{x}) - [\mathbf{f}(\mathbf{a}) + B(\mathbf{x} - \mathbf{a})]}{\|\mathbf{x} - \mathbf{a}\|}.$$

Subtracting these limits we have

$$\begin{aligned}\mathbf{0} &= \lim_{\mathbf{x}\to\mathbf{a}} \left(\frac{\mathbf{f}(\mathbf{x}) - \mathbf{f}(\mathbf{a}) - A(\mathbf{x} - \mathbf{a})}{\|\mathbf{x} - \mathbf{a}\|} - \frac{\mathbf{f}(\mathbf{x}) - \mathbf{f}(\mathbf{a}) - B(\mathbf{x} - \mathbf{a})}{\|\mathbf{x} - \mathbf{a}\|}\right) \\ &= \lim_{\mathbf{x}\to\mathbf{a}} \frac{(B - A)(\mathbf{x} - \mathbf{a})}{\|\mathbf{x} - \mathbf{a}\|}.\end{aligned}$$

(b) When taking the limit, it's possible to have $\mathbf{x} \to \mathbf{a}$ in a completely arbitrary manner. But one way to have $\mathbf{x} \to \mathbf{a}$ is along a straight-line path, which may be described as $\mathbf{x} = \mathbf{a} + t\mathbf{h}$. For such paths, having $\mathbf{x} \to \mathbf{a}$ is achieved by letting $t \to 0$. Thus if we know that

$$\lim_{\mathbf{x}\to\mathbf{a}} \frac{(B - A)(\mathbf{x} - \mathbf{a})}{\|\mathbf{x} - \mathbf{a}\|} = \mathbf{0},$$

then it must follow that

$$\lim_{t\to 0} \frac{(B - A)(t\mathbf{h})}{\|t\mathbf{h}\|} = \mathbf{0}.$$

(Note: The converse need *not* be true.) It follows that we must have a consistent one-sided limit; hence

$$\lim_{t\to 0^+} \frac{(B-A)(t\mathbf{h})}{\|t\mathbf{h}\|} = \mathbf{0}.$$

Now, for $t > 0$, we have

$$\frac{(B-A)(t\mathbf{h})}{\|t\mathbf{h}\|} = \frac{(B-A)\mathbf{h}}{\|\mathbf{h}\|}.$$

Thus if

$$\lim_{t\to 0^+} \frac{(B-A)(t\mathbf{h})}{\|t\mathbf{h}\|} = \lim_{t\to 0^+} \frac{(B-A)\mathbf{h}}{\|\mathbf{h}\|} = \mathbf{0},$$

it must be the case that $(B-A)\mathbf{h} = \mathbf{0}$. Moreover, this must be true for *any* nonzero vector $\mathbf{h} \in \mathbf{R}^n$. By setting $\mathbf{h}$ in turn equal to the standard basis vectors $\mathbf{e}_1, \ldots, \mathbf{e}_n$, we conclude that $B - A$ must be the zero matrix. Similarly, we must also have

$$\lim_{t\to 0^-} \frac{(B-A)(t\mathbf{h})}{\|t\mathbf{h}\|} = \mathbf{0}.$$

For $t < 0$, we have

$$\frac{(B-A)(t\mathbf{h})}{\|t\mathbf{h}\|} = \frac{(B-A)(t\mathbf{h})}{|t|\|\mathbf{h}\|} = -\frac{(B-A)\mathbf{h}}{\|\mathbf{h}\|}.$$

Thus if

$$\lim_{t\to 0^-} \frac{(B-A)(t\mathbf{h})}{\|t\mathbf{h}\|} = \lim_{t\to 0^-} -\frac{(B-A)\mathbf{h}}{\|\mathbf{h}\|} = \mathbf{0},$$

again it must be the case that $(B-A)\mathbf{h} = \mathbf{0}$. Hence $B - A$ must be the zero matrix.

2.4 Properties; Higher-Order Partial Derivatives

3. $\mathbf{f}(x,y,z) = (x \sin y + z, ye^x - 3x^2)$ and $\mathbf{g}(x,y,z) = (x^3 \cos x, xyz)$, so

$$D\mathbf{f} = \begin{bmatrix} \sin y & x\cos y & 1 \\ -6x & e^z & ye^z \end{bmatrix}, D\mathbf{g} = \begin{bmatrix} 3x^2\cos x - x^3 \sin x & 0 & 0 \\ yz & xz & xy \end{bmatrix} \quad \text{and}$$

$$D(\mathbf{f}+\mathbf{g}) = \begin{bmatrix} \sin y + 3x^2 \cos x - x^3 \sin x & x\cos y & 1 \\ -6x + yz & e^z + xz & ye^z + xy \end{bmatrix}.$$

7. $f(x,y) = 3xy + y^5, g(x,y) = x^3 - 2xy^2, f(x,y)g(x,y) = 3x^4y + x^3y^5 - 6x^2y^3 - 2xy^7$, and $\dfrac{f(x,y)}{g(x,y)} = \dfrac{3xy + y^5}{x^3 - 2xy^2}$. So

$$\begin{aligned} Df &= [3y, 3x + 5y^4], \quad \text{and} \quad Dg = [3x^2 - 2y^2, -4xy], \\ D(fg) &= [12x^3y + 3x^2y^5 - 12xy^3 - 2y^7, 3x^4 + 5x^3y^4 - 18x^2y^2 - 14xy^6] \\ &= (3xy + y^5)[3x^2 - 2y^2, -4xy] + (x^3 - 2xy^2)[3y, 3x + 5y^4] \\ &= fD(g) + gD(f), \text{and} \\ D\left(\frac{f}{g}\right) &= \left[\frac{g(x,y)f_x(x,y) - f(x,y)g_x(x,y)}{[g(x,y)]^2}, \frac{g(x,y)f_y(x,y) - f(x,y)g_y(x,y)}{[g(x,y)]^2}\right] \\ &= \frac{gDf - fDg}{g^2}. \end{aligned}$$

11. $f(x,y) = e^{y/x} - ye^{-x}$ so $f_x(x,y) = \dfrac{-y}{x^2}e^{y/x} + ye^{-x}$ and $f_y(x,y) = \dfrac{1}{x}e^{y/x} - e^{-x}$. The second order partials are:

$$\begin{aligned} f_{xx}(x,y) &= \frac{2y}{x^3}e^{y/x} + \frac{y^2}{x^4}e^{y/x} - ye^{-x}, \\ f_{xy}(x,y) = f_{yx}(x,y) &= \frac{-1}{x^2}e^{y/x} - \frac{y}{x^3}e^{y/x} + e^{-x}, \text{ and} \\ f_{yy}(x,y) &= \frac{1}{x^2}e^{y/x}. \end{aligned}$$

15. $f(x, y) = y \sin x - x \cos y$, so

$$f_x(x, y) = y \cos x - \cos y \quad \text{and} \quad f_y(x, y) = \sin x + x \sin y.$$

The second order partial derivatives are:

$$\begin{aligned} f_{xx}(x, y) &= -y \sin x, \\ f_{xy}(x, y) = f_{yx}(x, y) &= \cos x + \sin y, \quad \text{and} \\ f_{yy}(x, y) &= x \cos y. \end{aligned}$$

19. $f(x, y, z) = x^2yz + xy^2z + xyz^2$ so $f_x(x, y, z) = 2xyz + y^2z + yz^2$, $f_y(x, y, z) = x^2z + 2xyz + xz^2$, and $f_z(x, y, z) = x^2y + xy^2 + 2xyz$. The second order partials are:

$$\begin{aligned} f_{xx}(x, y, z) &= 2yz \\ f_{yy}(x, y, z) &= 2xz \\ f_{zz}(x, y, z) &= 2xy \\ f_{xy}(x, y, z) = f_{yx}(x, y, z) &= 2xz + 2yz + z^2 \\ f_{xz}(x, y, z) = f_{zx}(x, y, z) &= 2xy + y^2 + 2yz \\ f_{yz}(x, y, z) = f_{zy}(x, y, z) &= x^2 + 2xy + 2xz \end{aligned}$$

21. $f(x, y, z) = e^{ax} \sin y + e^{bx} \cos z$ so $f_x(x, y, z) = ae^{ax} \sin y + be^{bx} \cos z$, $f_y(x, y, z) = e^{ax} \cos y$, and $f_z(x, y, z) = -e^{bx} \sin z$. The second order partials are:

$$\begin{aligned} f_{xx}(x, y, z) &= a^2e^{ax} \sin y + b^2e^{bx} \cos z \\ f_{yy}(x, y, z) &= -e^{ax} \sin y \\ f_{zz}(x, y, z) &= -e^{bx} \cos z \\ f_{xy}(x, y, z) = f_{yx}(x, y, z) &= ae^{ax} \cos y \\ f_{xz}(x, y, z) = f_{zx}(x, y, z) &= -be^{bx} \sin z \\ f_{yz}(x, y, z) = f_{zy}(x, y, z) &= 0 \end{aligned}$$

25. First, for $f(x, y, z) = \ln\left(\frac{xy}{z}\right)$, we have

$$\begin{aligned} f_x(x, y, z) &= \left(\frac{z}{xy}\right)\left(\frac{y}{z}\right) = \frac{1}{x}, \\ f_y(x, y, z) &= \left(\frac{z}{xy}\right)\left(\frac{x}{z}\right) = \frac{1}{y}, \\ f_z(x, y, z) &= \left(\frac{z}{xy}\right)\left(-\frac{xy}{z^2}\right) = -\frac{1}{z}. \end{aligned}$$

From this, we see that, for $n \geq 1$.

$$\frac{\partial^n f}{\partial x^n} = \frac{(-1)^{n-1}(n-1)!}{x^n}, \quad \frac{\partial^n f}{\partial y^n} = \frac{(-1)^{n-1}(n-1)!}{y^n}, \quad \text{and} \quad \frac{\partial^n f}{\partial z^n} = \frac{(-1)^n(n-1)!}{z^n}.$$

Note that all mixed partials of this function are zero, since the first-order partial derivatives each involve just a single variable.

29. **(a)** To show that $T(x, t) = e^{-kt} \cos x$ satisfies the differential equation $kT_{xx} = T_t$ we calculate the derivatives:

$$\begin{aligned} T_x(x, t) &= -e^{-kt} \sin x \\ T_{xx}(x, t) &= -e^{-kt} \cos x \\ T_t(x, t) &= -ke^{-kt} \cos x \end{aligned}$$

so $kT_{xx} = T_t$.

For $t_0 = 0$ and $t_0 = 1$ the graphs are:

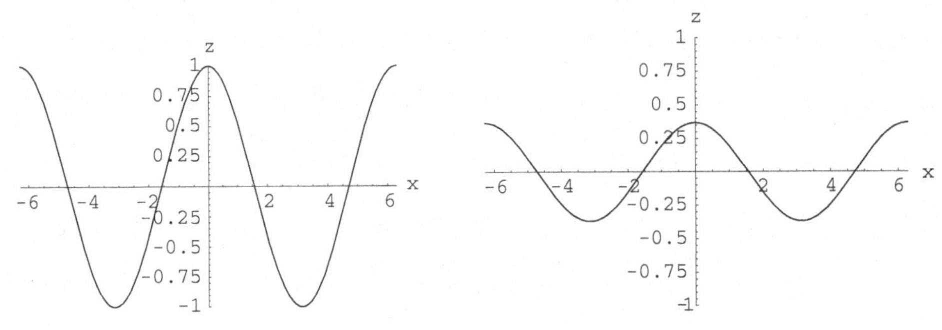

For $t_0 = 10$ the graph is further damped. The graph of the surface $z = T(x, t)$ is:

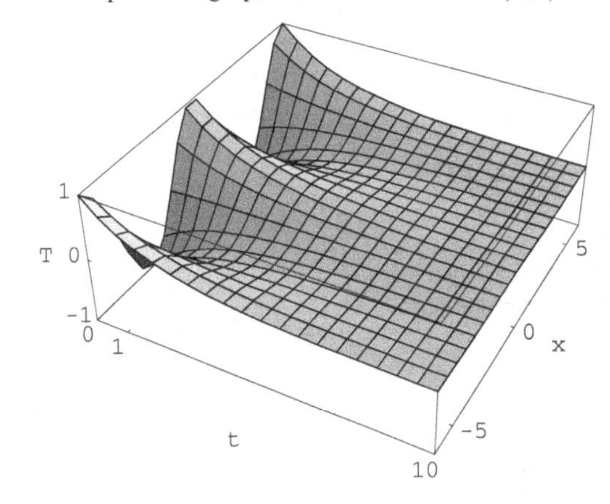

(b) To show that $T(x, y, t) = e^{-kt}(\cos x + \cos y)$ satisfies the differential equation $k(T_{xx} + T_{yy}) = T_t$ we calculate the derivatives:

$$T_x(x, y, t) = -e^{-kt} \sin x$$

$$T_{xx}(x, y, t) = -e^{-kt} \cos x$$

$$T_y(x, y, t) = -e^{-kt} \sin y$$

$$T_{yy}(x, y, t) = -e^{-kt} \cos y$$

$$T_t(x, y, t) = -ke^{-kt}(\cos x + \cos y)$$

so $k(T_{xx} + T_{yy}) = T_t$.

The graphs of the surfaces given by $z = T(x, y, t_0)$ for $t_0 = 0, 1$, and 10 are:

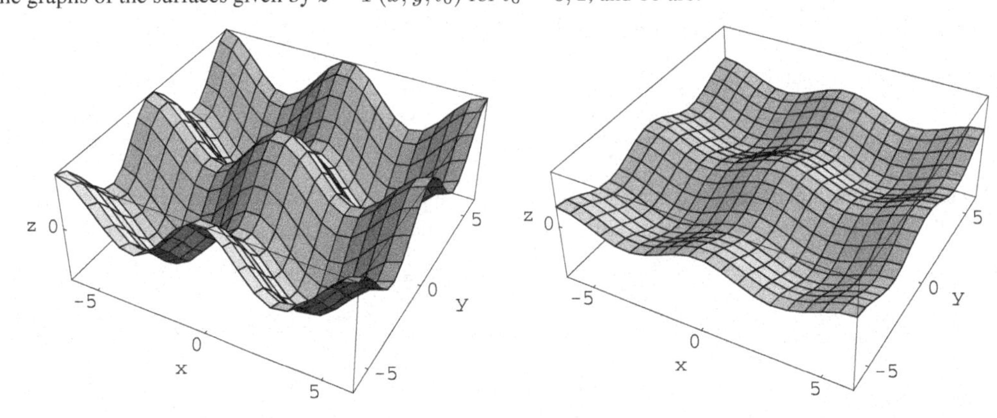

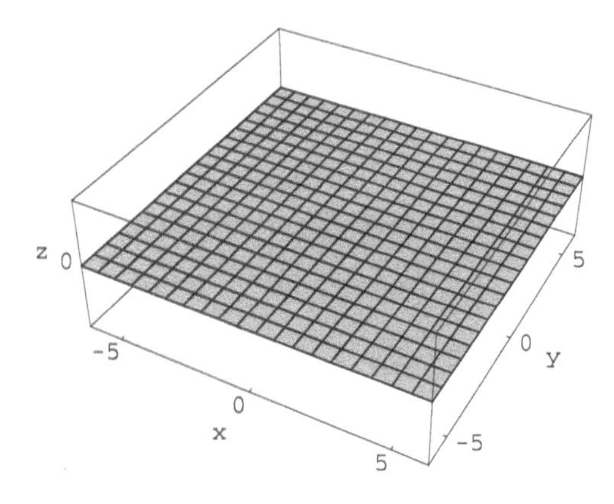

(c) Finally, to show that $T(x, y, z, t) = e^{-kt}(\cos x + \cos y + \cos z)$ satisfies the differential equation $k(T_{xx} + T_{yy} + T_{xx}) = T_t$ we calculate the derivatives:

$$\begin{aligned}
T_x(x, y, z, t) &= -e^{-kt} \sin x \\
T_{xx}(x, y, z, t) &= -e^{-kt} \cos x \\
T_y(x, y, z, t) &= -e^{-kt} \sin y \\
T_{yy}(x, y, z, t) &= -e^{-kt} \cos y \\
T_z(x, y, z, t) &= -e^{-kt} \sin z \\
T_{zz}(x, y, z, t) &= -e^{-kt} \cos z \\
T_t(x, y, z, t) &= -ke^{-kt}(\cos x + \cos y + \cos z)
\end{aligned}$$

so $k(T_{xx} + T_{yy} + T_{zz}) = T_t$.

33. (a) Here's an image of the helicoid:

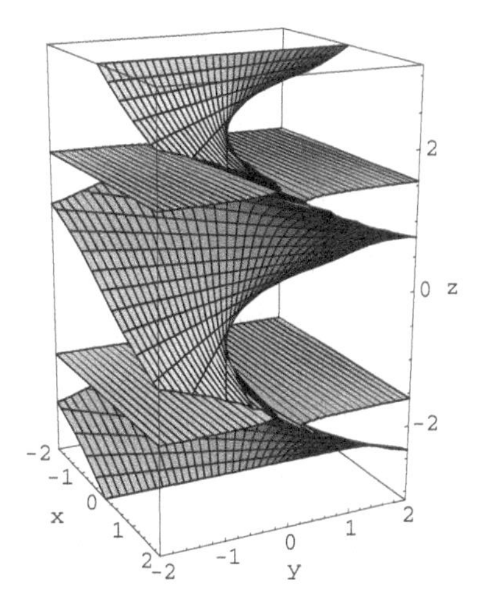

(b) There's no reason not to think of this surface as $z = x \tan y$. Then $z_x = \tan y$, $z_y = x \sec^2 y$, $z_{xx} = 0$, $z_{xy} = \sec^2 y$, and $z_{yy} = 2 \tan y \sec^2 y$. So

$$\begin{aligned}
(1 + z_y^2)z_{xx} + (1 + z_x^2)z_{yy} &= (1 + x^2 \sec^4 y)(0) + (1 + \tan^2 y)(2 \tan y \sec^2 y) \\
&= (\sec^2 y)(2 \tan y \sec^2 y) = 2(\tan y)(x \sec^2 y)(\sec^2 y) \\
&= 2z_x z_y z_{xy}
\end{aligned}$$

2.5 The Chain Rule

1. $f(x, y, z) = x^2 - y^3 + xyz$, $x = 6t + 7$, $y = \sin 2t$, and $z = t^2$.

Substitution:

$$\begin{aligned} f(x(t), y(t), z(t)) &= (6t+7)^2 - (\sin 2t)^3 + (6t+7)(\sin 2t)(t^2) \\ &= (6t+7)^2 - (\sin 2t)^3 + (6t^3 + 7t^2)(\sin 2t) \quad \text{and so} \\ \frac{df}{dt} &= 2(6t+7)6 - 3(\sin 2t)^2(2\cos 2t) + (18t^2 + 14t)\sin 2t + (6t^3 + 7t^2)(2\cos 2t) \end{aligned}$$

Chain Rule:

$$\begin{aligned} \frac{df}{dt} &= \frac{\partial f}{\partial x}\frac{dx}{dt} + \frac{\partial f}{\partial y}\frac{dy}{dt} + \frac{\partial f}{\partial z}\frac{dz}{dt} \\ &= (2x + yz)(6) + (-3y^2 + xz)(2\cos 2t) + (xy)(2t) \\ &= [2(6t+7) + (\sin 2t)(t^2)](6) + [-3\sin^2 2t + (6t+7)t^2](2\cos 2t) + [(6t+7)\sin 2t](2t) \end{aligned}$$

5. Here $V = LWH$, so

$$\begin{aligned} \frac{dV}{dt} &= \frac{\partial V}{\partial L}\frac{dL}{dt} + \frac{\partial V}{\partial W}\frac{dW}{dt} + \frac{\partial V}{\partial H}\frac{dH}{dt} \\ &= WH\left(\frac{dL}{dt}\right) + LH\left(\frac{dW}{dt}\right) + LW\left(\frac{dH}{dt}\right) \\ &= 5 \cdot 4(.75) + 7 \cdot 4(.5) + 7 \cdot 5(-1) \\ &= -6\ in^3/min. \end{aligned}$$

Since $\frac{dV}{dt} < 0$, the volume of the dough is decreasing at this instant.

7. Note that in 6 months:

$$\begin{aligned} x &= 1 + .6 - \cos\pi = 2.6 \\ y &= 200 + 12\sin\pi = 200 \end{aligned}$$

The chain rule gives

$$\begin{aligned} \left.\frac{dP}{dt}\right|_{t=6} &= \left.\frac{\partial P}{\partial x}\right|_{\substack{x=2.6\\y=200}} \left.\frac{dx}{dt}\right|_{t=6} + \left.\frac{\partial P}{\partial y}\right|_{\substack{x=2.6\\y=200}} \left.\frac{dy}{dt}\right|_{t=6} \\ &= 10(0.1x + 10)^{-\frac{1}{2}}(0.1)|_{x=2.6}\left.\left(0.1 - \frac{\pi}{6}\sin\frac{\pi t}{6}\right)\right|_{t=6} \\ &\quad - 4y^{-\frac{2}{3}}|_{y=200}\left.\left(2\sin\frac{\pi t}{6} + \frac{2\pi t}{6}\cos\frac{\pi t}{6}\right)\right|_{t=6} \\ &= (10.26)^{-\frac{1}{2}}(0.1) - 4(200^{-\frac{2}{3}})(-2\pi) \\ &= 0.031219527 + 0.734885812 = 0.766105339 \text{ units/month (demand in rising slightly).} \end{aligned}$$

11. Since $x = e^r\cos\theta$ and $y = e^r\sin\theta$ we can write

$$\frac{\partial z}{\partial r} = \left(\frac{\partial z}{\partial x}\right)\left(\frac{\partial x}{\partial r}\right) + \left(\frac{\partial z}{\partial y}\right)\left(\frac{\partial y}{\partial r}\right) = \left(\frac{\partial z}{\partial x}\right)(e^r\cos\theta) + \left(\frac{\partial z}{\partial y}\right)(e^r\sin\theta).$$

Similarly,

$$\frac{\partial z}{\partial \theta} = \left(\frac{\partial z}{\partial x}\right)\left(\frac{\partial x}{\partial \theta}\right) + \left(\frac{\partial z}{\partial y}\right)\left(\frac{\partial y}{\partial \theta}\right) = \left(\frac{\partial z}{\partial x}\right)(-e^r\sin\theta) + \left(\frac{\partial z}{\partial y}\right)(e^r\cos\theta).$$

Therefore,

$$\left(\frac{\partial z}{\partial r}\right)^2+\left(\frac{\partial z}{\partial \theta}\right)^2=e^{2r}\left[(\cos^2\theta+\sin^2\theta)\left(\frac{\partial z}{\partial x}\right)^2\right.$$

$$\left.+(\cos^2\theta+\sin^2\theta)\left(\frac{\partial z}{\partial y}\right)^2+(2\cos\theta\sin\theta-2\cos\theta\sin\theta)\left(\frac{\partial z}{\partial x}\right)\left(\frac{\partial z}{\partial y}\right)\right]$$

$$=e^{2r}\left[\left(\frac{\partial z}{\partial x}\right)^2+\left(\frac{\partial z}{\partial y}\right)^2\right].$$

The result follows.

15. First calculate:

$$\frac{\partial u}{\partial x}=\frac{y(y^2-x^2)}{(x^2+y^2)^2}\text{ and}$$

$$\frac{\partial u}{\partial y}=\frac{x(x^2-y^2)}{(x^2+y^2)^2}$$

Now

$$\begin{aligned}x\frac{\partial w}{\partial x}+y\frac{\partial w}{\partial y}&=x\frac{\partial w}{\partial u}\frac{\partial u}{\partial x}+y\frac{\partial w}{\partial u}\frac{\partial u}{\partial y}\\&=\left(\frac{\partial w}{\partial u}\right)\left(x\frac{\partial u}{\partial x}+y\frac{\partial u}{\partial y}\right)\\&=\left(\frac{\partial w}{\partial u}\right)\left(x\frac{y(y^2-x^2)}{(x^2+y^2)^2}+y\frac{x(x^2-y^2)}{(x^2+y^2)^2}\right)\\&=0.\end{aligned}$$

19. (a) $\mathbf{f}\circ g=(3(s-7t)^5,e^{2s-14t})$ so

$$D(\mathbf{f}\circ g)=\begin{bmatrix}15(s-7t)^4 & -105(s-7t)^4\\ 2e^{2s-14t} & -14e^{2x-14t}\end{bmatrix}$$

(b)

$$D\mathbf{f}=\begin{bmatrix}15x^4\\ 2e^{2x}\end{bmatrix}=\begin{bmatrix}15(s-7t)^4\\ 2e^{2s-14t}\end{bmatrix}\quad\text{and}\quad Dg=\begin{bmatrix}1 & -7\end{bmatrix}$$

We can easily see that $D\mathbf{f}\,Dg=D(\mathbf{f}\circ g)$.

23. (a) $\mathbf{f}\circ\mathbf{g}=\left((s/t)s^2t-\frac{s^2t}{s/t},\frac{s/t}{s^2t}+s^6t^3\right)=\left(s^3-st^2,\frac{1}{st^2}+s^6t^3\right)$, so

$$D(\mathbf{f}\circ\mathbf{g})=\begin{bmatrix}3s^2-t^2 & -2st\\ -1/(s^2t^2)+6s^5t^3 & -2/(st^3)+3s^6t^2\end{bmatrix}$$

(b)

$$D\mathbf{f}=\begin{bmatrix}y+\frac{y}{x^2} & x-\frac{1}{x}\\ \frac{1}{y} & \frac{-x}{y^2}+3y^2\end{bmatrix}=\begin{bmatrix}s^2t+\frac{s^2t}{s^2/t^2} & \frac{s}{t}-\frac{t}{s}\\ \frac{1}{s^2t} & \frac{-s/t}{s^4t^2}+3s^4t^2\end{bmatrix}=\begin{bmatrix}s^2t+t^3 & \frac{s^2-t^2}{st}\\ \frac{1}{s^2t} & -\frac{1}{s^3t^3}+3s^4t^2\end{bmatrix}$$

and $D\mathbf{g}=\begin{bmatrix}\frac{1}{t} & -\frac{s}{t^2}\\ 2st & s^2\end{bmatrix}$ so

$$D\mathbf{f}\,D\mathbf{g}=\begin{bmatrix}s^2t+t^3 & \frac{s^2-t^2}{st}\\ \frac{1}{s^2t} & -\frac{1}{s^3t^3}+3s^4t^2\end{bmatrix}\begin{bmatrix}\frac{1}{t} & -\frac{s}{t^2}\\ 2st & s^2\end{bmatrix}=\begin{bmatrix}3s^2-t^2 & -2st\\ \frac{-1}{s^2t^2}+6s^5t^3 & \frac{-2}{st^3}+3s^6t^2\end{bmatrix}.$$

27. (a) $\mathbf{f}\circ\mathbf{g}=(st+tu+su,s^3t^3-e^{stu^2})$ so

$$D(\mathbf{f}\circ\mathbf{g})=\begin{bmatrix}t+u & s+u & s+t\\ 3s^2t^3-tu^2e^{stu^2} & 3s^3t^2-su^2e^{stu^2} & -2stue^{stu^2}\end{bmatrix}.$$

(b)

$$D\mathbf{f} = \begin{bmatrix} 1 & 1 & 1 \\ 3x^2 & -ze^{yz} & -ye^{yz} \end{bmatrix} = \begin{bmatrix} 1 & 1 & 1 \\ 3s^2t^2 & -sue^{stu^2} & -tue^{stu^2} \end{bmatrix}$$

and $D\mathbf{g} = \begin{bmatrix} t & s & 0 \\ 0 & u & t \\ u & 0 & s \end{bmatrix}$ so $D\mathbf{f}D\mathbf{g} = \begin{bmatrix} t+u & s+u & s+t \\ 3s^2t^3 - tu^2e^{stu^2} & 3s^3t^2 - su^2e^{stu^2} & -2stue^{stu^2} \end{bmatrix}.$

31. (a) From formula (10) in Section 2.5, we have

$$\frac{\partial}{\partial x} = \cos\theta\frac{\partial}{\partial r} - \frac{\sin\theta}{r}\frac{\partial}{\partial \theta} \text{ and}$$
$$\frac{\partial}{\partial y} = \sin\theta\frac{\partial}{\partial r} + \frac{\cos\theta}{r}\frac{\partial}{\partial \theta}$$

Hence if $z = f(x, y)$, then

$$\frac{\partial^2 z}{\partial x^2} = \frac{\partial}{\partial x}\left(\frac{\partial z}{\partial x}\right) = \cos\theta\frac{\partial}{\partial r}\left(\frac{\partial z}{\partial x}\right) - \frac{\sin\theta}{r}\frac{\partial}{\partial\theta}\left(\frac{\partial z}{\partial x}\right)$$
$$= \cos\theta\frac{\partial}{\partial r}\left(\cos\theta\frac{\partial z}{\partial r} - \frac{\sin\theta}{r}\frac{\partial z}{\partial\theta}\right) - \frac{\sin\theta}{r}\frac{\partial}{\partial\theta}\left(\cos\theta\frac{\partial z}{\partial r} - \frac{\sin\theta}{r}\frac{\partial z}{\partial\theta}\right)$$

Now use the product rule:

$$\frac{\partial^2 z}{\partial x^2} = \cos\theta\left(\cos\theta\frac{\partial^2 z}{\partial r^2} + \frac{\sin\theta}{r^2}\frac{\partial z}{\partial\theta} - \frac{\sin\theta}{r}\frac{\partial^2 z}{\partial r\partial\theta}\right)$$
$$- \frac{\sin\theta}{r}\left(-\sin\theta\frac{\partial z}{\partial r} + \cos\theta\frac{\partial^2 z}{\partial\theta\partial r} - \frac{\cos\theta}{r}\frac{\partial z}{\partial\theta} - \frac{\sin\theta}{r}\frac{\partial^2 z}{\partial\theta^2}\right)$$
$$= \cos^2\theta\frac{\partial^2 z}{\partial r^2} + \frac{2\sin\theta\cos\theta}{r^2}\frac{\partial z}{\partial\theta} - \frac{2\sin\theta\cos\theta}{r}\frac{\partial^2 z}{\partial r\partial\theta} + \frac{\sin^2\theta}{r}\frac{\partial z}{\partial r} + \frac{\sin^2\theta}{r^2}\frac{\partial^2 z}{\partial\theta^2}.$$

Follow the same steps to calculate

$$\frac{\partial^2 z}{\partial y^2} = \frac{\partial}{\partial y}\left(\frac{\partial z}{\partial y}\right) = \sin\theta\frac{\partial}{\partial r}\left(\frac{\partial z}{\partial y}\right) + \frac{\cos\theta}{r}\frac{\partial}{\partial\theta}\left(\frac{\partial z}{\partial y}\right)$$
$$= \sin\theta\frac{\partial}{\partial r}\left(\sin\theta\frac{\partial z}{\partial r} + \frac{\cos\theta}{r}\frac{\partial z}{\partial\theta}\right) + \frac{\cos\theta}{r}\frac{\partial}{\partial\theta}\left(\sin\theta\frac{\partial z}{\partial r} + \frac{\cos\theta}{r}\frac{\partial z}{\partial\theta}\right)$$
$$= \sin^2\theta\frac{\partial^2 z}{\partial r^2} - \frac{2\sin\theta\cos\theta}{r^2}\frac{\partial z}{\partial\theta} + \frac{2\sin\theta\cos\theta}{r}\frac{\partial^2 z}{\partial\theta\partial r} + \frac{\cos^2\theta}{r}\frac{\partial z}{\partial r} + \frac{\cos^2\theta}{r^2}\frac{\partial^2 z}{\partial\theta^2}.$$

(b) Adding the two equations above we easily see that

$$\frac{\partial^2}{\partial x^2} + \frac{\partial^2}{\partial y^2} = \frac{\partial^2}{\partial r^2} + \frac{1}{r}\frac{\partial}{\partial r} + \frac{1}{r^2}\frac{\partial^2}{\partial\theta^2}.$$

33. (a) The chain rule gives $\frac{\partial w}{\partial\rho} = \frac{\partial w}{\partial r}\frac{\partial r}{\partial\rho} + \frac{\partial w}{\partial\theta}\frac{\partial\theta}{\partial\rho} + \frac{\partial w}{\partial z}\frac{\partial z}{\partial\rho}$ for any appropriately differentiable function w. Now (6) of §1.7 gives $z = \rho\cos\varphi, r = \rho\sin\varphi$. Hence

$$\frac{\partial w}{\partial\rho} = \sin\varphi\frac{\partial w}{\partial r} + 0 + \cos\varphi\frac{\partial w}{\partial z} = \sin\varphi\frac{\partial w}{\partial r} + \cos\varphi\frac{\partial w}{\partial z}.$$

Also

$$\frac{\partial w}{\partial\varphi} = \rho\cos\varphi\frac{\partial w}{\partial r} - \rho\sin\varphi\frac{\partial w}{\partial z} \quad \text{from a similar chain rule computation.}$$

From this, we have

$$\rho\sin\varphi\frac{\partial w}{\partial\rho} + \cos\varphi\frac{\partial w}{\partial\varphi} = \left(\rho\sin^2\varphi\frac{\partial w}{\partial r} + \rho\sin\varphi\cos\varphi\frac{\partial w}{\partial z}\right) + \left(\rho\cos^2\varphi\frac{\partial w}{\partial r} - \rho\cos\varphi\sin\varphi\frac{\partial w}{\partial z}\right)$$
$$= \rho\frac{\partial w}{\partial r}.$$

Thus

$$\frac{\partial w}{\partial r} = \sin\varphi \frac{\partial w}{\partial \rho} + \frac{\cos\varphi}{\rho}\frac{\partial w}{\partial \varphi} \quad \text{or} \quad \frac{\partial}{\partial r} = \sin\varphi \frac{\partial}{\partial \rho} + \frac{\cos\varphi}{\rho}\frac{\partial}{\partial \varphi}.$$

(Alternatively, consider formula (10) in this section with $x = z$, $y = r$, θ replaced by φ, and r replaced by ρ.)

(b) The cylindrical Laplacian is $\frac{\partial^2}{\partial r^2} + \frac{\partial^2}{\partial z^2} + \frac{1}{r^2}\frac{\partial^2}{\partial \theta^2} + \frac{1}{r}\frac{\partial}{\partial r}$. From $z = \rho\cos\varphi$, $r = \rho\sin\varphi$, we may treat z and r as if they are Cartesian coordinates, so that

$$\frac{\partial^2}{\partial r^2} + \frac{\partial^2}{\partial z^2} = \frac{\partial^2}{\partial \rho^2} + \frac{1}{\rho^2}\frac{\partial^2}{\partial \varphi^2} + \frac{1}{\rho}\frac{\partial}{\partial \rho} \quad \text{(Cartesian/}\textit{cylindrical}\text{)}$$

Now we know $\frac{\partial}{\partial r}$ from part (a). So, with $r = \rho\sin\varphi$, we have

$$\begin{aligned}\left(\frac{\partial^2}{\partial r^2} + \frac{\partial^2}{\partial z^2}\right) + \frac{1}{r^2}\frac{\partial^2}{\partial \theta^2} + \frac{1}{r}\frac{\partial}{\partial r} &= \left(\frac{\partial^2}{\partial \rho^2} + \frac{1}{\rho^2}\frac{\partial^2}{\partial \varphi^2} + \frac{1}{\rho}\frac{\partial}{\partial \rho}\right) \\ &\quad + \frac{1}{\rho^2\sin^2\varphi}\frac{\partial^2}{\partial \theta^2} + \frac{1}{\rho\sin\varphi}\left(\sin\varphi\frac{\partial}{\partial \rho} + \frac{\cos\varphi}{\rho}\frac{\partial}{\partial \varphi}\right) \\ &= \frac{\partial^2}{\partial \rho^2} + \frac{1}{\rho^2}\frac{\partial^2}{\partial \varphi^2} + \frac{2}{\rho}\frac{\partial}{\partial \rho} + \frac{1}{\rho^2\sin^2\varphi}\frac{\partial^2}{\partial \theta^2} + \frac{\cot\varphi}{\rho^2}\frac{\partial}{\partial \varphi} \quad \text{as desired.}\end{aligned}$$

37. Use the equations from Exercise 36(a) for $F(x, y, z) = x^3 z + y\cos z + (\sin y)/z = 0$:

$$\frac{\partial z}{\partial x} = \frac{-3x^2 z}{x^3 - y\sin z - (\sin y)/z^2} = \frac{-3x^2 z^3}{x^3 z^2 - yz^2\sin z - \sin y} \quad \text{and}$$

$$\frac{\partial z}{\partial y} = \frac{-\cos z - (\cos y)/z}{x^3 - y\sin z - (\sin y)/z^2} = \frac{-z^2\cos z - z\cos y}{x^3 z^2 - yz^2\sin z - \sin y}.$$

41. $\left(\frac{\partial s}{\partial z}\right)_{x,y,w} = xw - 2z$, so $\left(\frac{\partial s}{\partial z}\right)_{x,w} = \left(\frac{\partial s}{\partial z}\right)_{x,y,w} + \left(\frac{\partial s}{\partial y}\right)_{x,z,w}\left(\frac{\partial y}{\partial z}\right)_{x,w}$.

To calculate $\left(\frac{\partial y}{\partial z}\right)_{x,w}$ we can use the results of Exercise 36 with $F(x, y, z, w) = xyw - y^3 z + xz$:

$$\left(\frac{\partial y}{\partial z}\right)_{x,w} = -\frac{F_z(x, y, z, w)}{F_y(x, y, z, w)} = -\frac{-y^3 + x}{xw - 3y^2 z}.$$

So $\left(\frac{\partial s}{\partial z}\right)_{x,w} = xw - 2z + (x^2)\left(\frac{y^3 - x}{xw - 3y^2}\right)$.

45. It is easiest to use implicit differentiation and solve. For example, for the equation $ax^2 + by^2 + cz^2 - d = 0$, hold z constant and take the derivative with respect to y. You get $2ax(\partial x/\partial y)_z + 2by = 0$. Solve this and get $(\partial x/\partial y)_z = -by/ax$. Similarly we get that $(\partial y/\partial z)_x = -cz/by$ and $(\partial z/\partial x)_y = -ax/cz$. So

$$\left(\frac{\partial x}{\partial y}\right)_z\left(\frac{\partial y}{\partial z}\right)_x\left(\frac{\partial z}{\partial x}\right)_y = \left(\frac{-by}{ax}\right)\left(\frac{-cz}{by}\right)\left(\frac{-ax}{cz}\right) = -1.$$

2.6 Directional Derivatives and the Gradient

1. **(a)** $\nabla f(x, y, z) \cdot (-\mathbf{k})$ is the directional derivative of $f(x, y, z)$ in the direction $-\mathbf{k}$ (i.e., the negative z direction).

(b) $\nabla f(x, y, z) \cdot (-\mathbf{k}) = \left(\frac{\partial f}{\partial x}, \frac{\partial f}{\partial y}, \frac{\partial f}{\partial z}\right) \cdot (0, 0, -1) = -\frac{\partial f}{\partial z}$.

5. $\nabla f(x, y) = (e^x - 2x, 0)$ so $\nabla f(1, 2) = (e - 2, 0)$ and

$$D_{\mathbf{u}} f(\mathbf{a}) = (e - 2, 0) \cdot \frac{(2, 1)}{\sqrt{5}} = \frac{2e - 4}{\sqrt{5}}.$$

11. The gradient direction for the function h is $\nabla h = (-6xy^2, -6x^2 y)$.

(a) Head in the direction $\nabla h(1,-2) = (-24, 12)$. If you prefer your directions given by a unit vector, we normalize to obtain:

$$\frac{\nabla h(1,-2)}{\|\nabla h(1,-2)\|} = \frac{(-24,12)}{\sqrt{24^2+12^2}} = \frac{(-2,1)}{\sqrt{5}}.$$

(b) Head in a direction orthogonal to your answer for part (a): $\pm\frac{(1,2)}{\sqrt{5}}$.

15. We want to head in the direction of the negative gradient. Since $M(x,y) = 3x^2 + y^2 + 5000$, the negative gradient is $-\nabla M(x,y) = (-6x, -2y)$. This means that

$$\frac{dy}{dx} = \frac{-2y}{-6x} = \frac{y}{3x}.$$

This is the separable differential equation $(3/y)\,dy = (1/x)\,dx$ or $3\ln y = \ln x + c$. Work the usual magic and get $y^3 = kx$. Substitute in the point (8, 6) to solve for k to end up with the path $y^3 = 27x$.

19. $f(x,y,z) = 2xy^2 - 2z^2 + xyz$ so $\nabla f(x,y,z) = (2y^2 + yz, 4xy + xz, xy - 4z)$ and $\nabla f(2,-3,3) = (9,-18,-18)$. So the equation of the tangent plane is:

$$0 = (9,-18,-18)\cdot(x-2, y+3, z-3) \quad \text{or} \quad x - 2y - 2z = 2.$$

23. The tangent plane to the surface at a point (x_0, y_0, z_0) is

$$0 = 18x_0(x-x_0) - 90y_0(y-y_0) + 10z_0(z-z_0).$$

For this to be parallel to $x + 5y - 2z = 7$, the vector

$$(18x_0, -90y_0, 10z_0) = k(1, 5, -2).$$

This means that $y_0 = -x_0$ and $z_0 = (-18/5)x_0$. Substitute these back into the equation of the hyperboloid: $9x^2 - 45y^2 + 5z^2 = 45$ to get:

$$45 = 9x_0^2 - 45x_0^2 + 5(18^2/5^2)x_0^2 \quad \text{therefore} \quad x_0 = \pm 5/4.$$

This means that the points are $(5/4, -5/4, -9/2)$ and $(-5/4, 5/4, 9/2)$.

27. (a) For $f(x,y,z) = x^3 - x^2y^2 + z^2$, $\nabla f(x,y,z) = (3x^2 - 2xy^2, -2x^2y, 2z)$ so $\nabla f(2,-3/2,1) = (3,12,2)$. Thus the equation of the tangent plane is

$$3(x-2) + 12(y+3/2) + 2(z-1) = 0 \quad \text{or} \quad 3x + 12y + 2z + 10 = 0.$$

(b) $\nabla f(0,0,0) = (0,0,0)$ so the gradient cannot be used as a normal vector. If we solve $z = \pm\sqrt{y^2x^2 - x^3} = \pm x\sqrt{y^2 - x}$, we see that $g(x,y) = x\sqrt{y^2 - x}$ fails to be differentiable at (0, 0)—so there is no tangent plane there.

31. If $f(x,y) = x^2 - y^2$ then $\nabla f(5,-4) = (10, 8)$ so the equations of the normal line are

$$x(t) = 10t + 5 \quad \text{and} \quad y(t) = 8t - 4 \quad \text{or} \quad 8x - 10y = 80.$$

35. Using the method above for $f(x,y,z) = e^{xy} + e^{zx} - 2e^{yz}$, we find that $\nabla f = (ye^{xy} + ze^{xz}, xe^{xy} - 2ze^{yz}, xe^{xz} - 2ye^{yz})$ so $\nabla f(-1,-1,-1) = e(-2,1,1)$. So

$$\begin{cases} x = -2et - 1 \\ y = et - 1 \\ z = et - 1 \end{cases} \quad \text{or, factoring out } e, \quad \begin{cases} x = -2t - 1 \\ y = t - 1 \\ z = t - 1. \end{cases}$$

39. Here $f(x_1, x_2, \ldots, x_n) = x_1^2 + x_2^2 + \cdots + x_n^2$ so $\nabla f(x_1, x_2, \ldots, x_n) = (2x_1, 2x_2, \ldots, 2x_n)$. Using the techniques of this section, the tangent hyperplane to the $(n-1)$-dimensional sphere $f(x_1, x_2, \ldots, x_n) = 1$ at $(1/\sqrt{n}, 1/\sqrt{n}, \ldots, 1/\sqrt{n}, -1/\sqrt{n})$ is

$$\begin{aligned} 0 &= \nabla f(1/\sqrt{n}, \ldots, 1/\sqrt{n}, -1/\sqrt{n})\cdot(x_1 - 1/\sqrt{n}, x_2 - 1/\sqrt{n}, \ldots, x_{n-1} - 1/\sqrt{n}, x_n + 1/\sqrt{n}) \\ &= \frac{2}{\sqrt{n}}\left(x_1 - \frac{1}{\sqrt{n}}\right) + \frac{2}{\sqrt{n}}\left(x_2 - \frac{1}{\sqrt{n}}\right) + \cdots + \frac{2}{\sqrt{n}}\left(x_{n-1} - \frac{1}{\sqrt{n}}\right) + \frac{-2}{\sqrt{n}}\left(x_n + \frac{1}{\sqrt{n}}\right) \quad \text{or} \end{aligned}$$

$$0 = (x_1 - 1/\sqrt{n}) + (x_2 - 1/\sqrt{n}) + \cdots + (x_{n-1} - 1/\sqrt{n}) - (x_n + 1/\sqrt{n}) \quad \text{so}$$

$$\sqrt{n} = x_1 + x_2 + \cdots + x_{n-1} - x_n.$$

41. (a) $\dfrac{\partial F}{\partial z} = xe^{xz}$. This is non-zero whenever $x \neq 0$. There we can solve for z to get

$$z = \frac{\ln(1 - \sin xy - x^3 y)}{x}.$$

(b) Looking only at points in S we only need to stay away from points in yz-plane (i.e., where $x = 0$).

(c) You shouldn't then make the leap from your answer to part (b) that you can graph $z = \dfrac{\ln(1 - \sin xy - x^3 y)}{x}$ for any values of x and y just so $x \neq 0$. Your other restriction is that $1 - \sin xy - x^3 y > 0$ as it is the argument of the natural logarithm. A sketch that gives you an idea of the surface is:

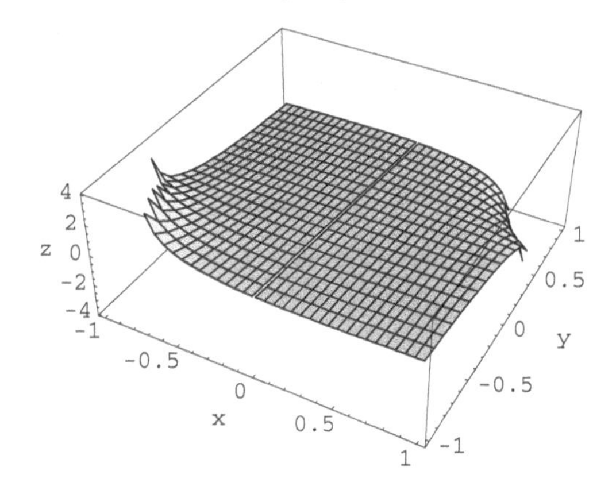

Now the actual surface S includes the plane $x = 0$ since $x = 0$ satisfies the original equation: $\sin xy + e^{xz} + x^3 y = 1$. S will actually look a bit like:

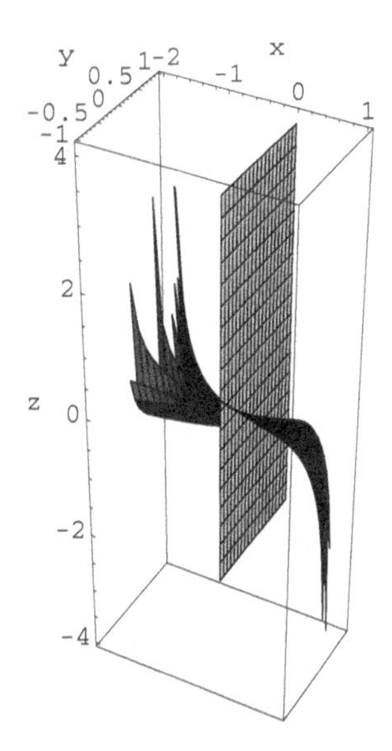

45. (a) $G(-1,1,1) = F(-1-2+1, -1-1+3) = F(-2,1) = 0$.

(b) To invoke the implicit function theorem, we need to show that $G_z(-1,1,1) \neq 0$.

$$\begin{aligned} G_z(-1,1,1) &= F_u(-2,1)\frac{\partial(x^3 - 2y^2 + z^5)}{\partial z}\bigg|_{(-1,1,1)} + F_v(-2,1)\frac{\partial(xy - x^2 z + 3)}{\partial z}\bigg|_{(-1,1,1)} \\ &= (7)(5) + (5)(1) = 40 \neq 0. \end{aligned}$$

49. **(a)** We need to consider where the same determinant is non-zero. In this case the determinant is

$$\begin{vmatrix} \sin\varphi\cos\theta & \rho\cos\varphi\cos\theta & -\rho\sin\varphi\sin\theta \\ \sin\varphi\sin\theta & \rho\cos\varphi\sin\theta & \rho\sin\varphi\cos\theta \\ \cos\varphi & -\rho\sin\varphi & 0 \end{vmatrix} = \rho^2\sin\varphi\cos^2\varphi + \rho^2\sin^3\varphi = \rho^2\sin\varphi.$$

In other words, for any points for which $\rho \neq 0$ and for which $\sin\varphi \neq 0$.

(b) Again, this makes complete sense. When the radius is 0, then ρ completely determines the point as being the origin. When $\sin\varphi = 0$ you are on the z-axis so θ no longer contributes any information.

2.7 Newton's Method

1. We begin by defining the function $\mathbf{f}(x, y) = (y^2e^x - 3, 2ye^x + 10y^4)$. Then we have

$$D\mathbf{f}(x, y) = \begin{bmatrix} y^2e^x & 2ye^x \\ 2ye^x & 2e^x + 40y^3 \end{bmatrix}.$$

The inverse of this matrix is

$$[D\mathbf{f}(x, y)]^{-1} = \begin{bmatrix} \dfrac{2e^x + 40y^3}{40y^5e^x - 2y^2e^{2x}} & \dfrac{1}{ye^x - 20y^4} \\ \dfrac{1}{ye^x - 20y^4} & \dfrac{1}{40y^3 - 2e^x} \end{bmatrix}.$$

Hence the iteration expression

$$\mathbf{x}_k = \mathbf{x}_{k-1} - [D\mathbf{f}(\mathbf{x}_{k-1})]^{-1}\mathbf{f}(\mathbf{x}_{k-1})$$

becomes (after some simplification using *Mathematica*)

$$x_k = \frac{x_{k-1}y_{k-1}^2e^{x_{k-1}} - y_{k-1}^2e^{x_{k-1}} + 10y_{k-1}^5 - 60e^{-x_{k-1}}y_{k-1}^3 - 20x_{k-1}y_{k-1}^5 - 3}{y_{k-1}^2e^{x_{k-1}} - 20y_{k-1}^5}$$

$$y_k = \frac{y_{k-1}^2e^{x_{k-1}} - 15y_{k-1}^5 + 3}{y_{k-1}e^{x_{k-1}} - 20y_{k-1}^4}.$$

Using initial vector $(x_0, y_0) = (1, -1)$ and iterating the formulas above we obtain the following results:

k	x_k	y_k
0	1	-1
1	1.279707977	-0.911965173
2	1.302659547	-0.902966291
3	1.302942519	-0.902880458
4	1.302942538	-0.902880451
5	1.302942538	-0.902880451

Since the result appears to be stable to nine decimal places, we conclude that the approximate solution is $(1.302942538, -0.902880451)$.

3. **(a)** The graphs of the curves are as follows:

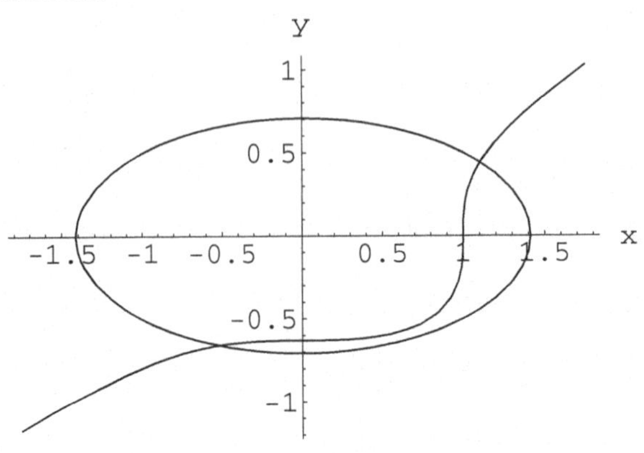

From the graph, we estimate one intersection point near $(1, 1/2)$, and a second near $(-1/2, -3/4)$.

(b) Using the function $\mathbf{f}(x, y) = (x^3 - 4y^3 - 1, x^2 + 4y^2 - 2)$, we have

$$D\mathbf{f}(x, y) = \begin{bmatrix} 3x^2 & -12y^2 \\ 2x & 8y \end{bmatrix} \implies [D\mathbf{f}(x, y)]^{-1} = \begin{bmatrix} \dfrac{1}{3x^2 + 3xy} & \dfrac{y}{2x^2 + 2xy} \\ -\dfrac{1}{12xy + 12y^2} & \dfrac{x}{8xy + 8y^2} \end{bmatrix}.$$

Then the iteration expression in formula (6) becomes

$$x_k = \frac{4x_{k-1}^3 + 3x_{k-1}^2 y_{k-1} - 4y_{k-1}^3 + 6y_{k-1} + 2}{6x_{k-1}(x_{k-1} + y_{k-1})}$$

$$y_k = \frac{-x_{k-1}^3 + 12x_{k-1}y_{k-1}^2 + 16y_{k-1}^3 + 6x_{k-1} - 2}{24y_{k-1}(x_{k-1} + y_{k-1})}.$$

Using initial vector $(x_0, y_0) = (1, 0.5)$ and iterating the formulas above we obtain the following results:

k	x_k	y_k
0	1	0.5
1	1.111111111	0.444444444
2	1.103968254	0.441964286
3	1.103931712	0.441965716
4	1.103931711	0.441965716
5	1.103931711	0.441965716

The data imply that to nine decimal places there is an intersection point at $(1.103931711, 0.441965716)$. Using initial vector $(x_0, y_0) = (-0.5, -0.75)$ and iterating, we find

k	x_k	y_k
0	-0.5	-0.75
1	-0.5	-0.666666667
2	-0.518518519	-0.657986111
3	-0.518214436	-0.65792361
4	-0.518214315	-0.657923613
5	-0.518214315	-0.657923613

Thus it appears that to nine decimal places there is an second intersection point at $(-0.518214315, -0.657923613)$.

7. Formula (6) says $\mathbf{x}_{k+1} = \mathbf{x}_k - [D\mathbf{f}(\mathbf{x}_k)]^{-1}\mathbf{f}(\mathbf{x}_k)$. But if $\mathbf{x}_k$ solves (2) exactly, then $\mathbf{f}(\mathbf{x}_k) = \mathbf{0}$. Thus $\mathbf{x}_{k+1} = \mathbf{x}_k - [D\mathbf{f}(\mathbf{x}_k)]^{-1}\mathbf{0} = \mathbf{x}_k$. By the same argument $\mathbf{x}_k = \mathbf{x}_{k+2} = \mathbf{x}_{k+3} = \cdots$.

True/False Exercises for Chapter 2

1. False.
3. False. (The range also requires $v \neq 0$.)
5. True.
7. False. (The graph of $x^2 + y^2 + z^2 = 0$ is a single point.)
9. False.
11. False. ($\lim_{(x,y)\to(0,0)} f(x, y) = 0 \neq 2$.)
13. False.
15. False. ($\nabla f(x, y, z) = (0, \cos y, 0)$.)
17. True.
19. False. (The partial derivatives must be continuous.)
21. False. ($f_{xy} \neq f_{yx}$.)
23. True. (Write the chain rule for this situation.)
25. False. (The correct equation is $4x + y + 4z = 0$.)
27. True.
29. False.

Miscellaneous Exercises for Chapter 2

1. (a) Calculate the determinant

$$\begin{vmatrix} \mathbf{i} & \mathbf{j} & \mathbf{k} \\ 1 & 0 & 1 \\ x_1 & x_2 & x_3 \end{vmatrix} = (-x_2, x_1 - x_3, x_2).$$

More explicitly, the component functions are $f_1(x_1, x_2, x_3) = -x_2$, $f_2(x_1, x_2, x_3) = x_1 - x_3$, and $f_3(x_1, x_2, x_3) = x_2$.

(b) The domain is all of $\mathbf{R}^3$ while the range restricts the first component to be the opposite of the last component. In other words the range is the set of all vectors $(a, b, -a)$.

5.

$f(x, y)$	Graph	Level curves
$1/(x^2 + y^2 + 1)$	D	d
$\sin\sqrt{x^2 + y^2}$	B	e
$(3y^2 - 2x^2)e^{-x^2-2y^2}$	A	b
$y^3 - 3x^2y$	E	c
$x^2y^2e^{-x^2-y^2}$	F	a
$ye^{-x^2-y^2}$	C	f

9. $g(x) = F(x, 0) \equiv 0$ and so is continuous at $x = 0$ and $h(y) = F(0, y) \equiv 0$ and so is continuous at $y = 0$. Consider $p(x) = F(x, x) = 1$ when $x \neq 0$ and $F(0, 0) = 0$. Clearly, $p(x)$ is not continuous at 0 so $F(x, y)$ is not continuous at $(0, 0)$.

15. (a) Comparison with Exercise 11: With an air temperature of 25° F, windspeed of 10 mph,

$$W(10, 25) = 91.4 + (25 - 91.4)(0.474 + 0.304\sqrt{10} - 0.203)$$
$$\approx 9.573 \quad \text{or} \quad 10^\circ \text{ F}$$

(as compared to 15° F in 11(a)).
If $s = 20$ mph, then $W = -15°$ F if

$$91.4 + (t - 91.4)(0.474 + 0.304\sqrt{20} - 0.406) = -15.$$

Hence $t = 91.4 - \dfrac{15 + 91.4}{(0.474 + 0.304\sqrt{20} - 0.406)} \approx 16.866$ or 17° F (as compared to 5° F in 11(b)).

Comparison with Exercise 12: With $W(s, t) = 91.4 + (t - 91.4)(.474 + .304\sqrt{s} - .0203s)$, we must solve

$$-5 = 91.4 + (10 - 91.4)(.474 + .304\sqrt{s} - .0203s)$$

or

$$\frac{-5 - 91.4}{10 - 91.4} = .474 + .304\sqrt{s} - .0203s \quad \text{so that}$$
$$1.18428 \approx .474 + .304\sqrt{s} - .0203s \quad \text{or}$$
$$0 \approx .0203s - .304\sqrt{s} + .7102752$$

Now solve the quadratic: $\sqrt{s} \approx \dfrac{.304 \pm \sqrt{.186391}}{.0406}$. The two solutions are 8.39128 and 145.893.

(b) The windchill effect of windspeed appears to be greater in the Siple formula than that which may be inferred from the table.

(c) For temperatures greater than 91.4 the model has the wind actually making the apparent temperature warmer than air temperature. Physically, the model probably falls apart because between 91.4 and 106 you are too close to body temperature for the wind to have much effect and if you are in temperatures much greater than 106 a breeze won't replace a frosty beverage. For winds below 4 mph, the effect is negligible and won't be reflected in the model.

19. Without loss of generality we can locate the center of the sphere at the origin and so the equation of the sphere is $F(x, y, z) = x^2 + y^2 + z^2 = r^2$ so $\nabla F = (2x, 2y, 2z)$ and the equation of the plane tangent to the sphere at $P = (x_0, y_0, z_0)$ is $0 = (2x_0, 2y_0, 2z_0) \cdot (x - x_0, y - y_0, z - z_0)$ or $x_0x + y_0y + z_0z = x_0^2 + y_0^2 + z_0^2$. This is orthogonal to the vector (x_0, y_0, z_0) which is the vector from the center of the sphere to P.

23. (a) As we have done before we find the plane tangent to the surface given by $F(x, y, z) = z - x^2 - 4y^2 = 0$ by formula (6):

$$0 = \nabla F(1, -1, 5) \cdot (x - 1, y + 1, z - 5) = (-2, 8, 1) \cdot (x - 1, y + 1, z - 5)$$
$$\text{or } -2x + 8y + z = -5.$$

(b) The line is parallel to a vector which is orthogonal to $\nabla F(1, -1, 5) = (-2, 8, 1)$ and with no component in the x direction. So it is of the form $(0, a, b)$ with $(0, a, b) \cdot (-2, 8, 1) = 0$ so the line has the direction $(0, 1, -8)$ and passes through $(1, -1, 5)$. The equations are $\begin{cases} x = 1 \\ y = t - 1 \\ z = -8t + 5. \end{cases}$

25. First note that $0.2\,\text{deg C/day} = 0.2 \cdot 24 = 4.8\,\text{deg C/month}$. Then, with time measured in months, the chain rule tells us

$$\frac{dP}{dt} = \frac{\partial P}{\partial S}\frac{dS}{dt} + \frac{\partial P}{\partial T}\frac{dT}{dt}.$$

Here $\frac{dS}{dt} = -2$, $\frac{dT}{dt} = 4.8$. With $P(S, T) = 330S^{2/3}T^{4/5}$, we have

$$\begin{aligned}
\left.\frac{dP}{dt}\right|_{(S=75, T=15)} &= (220S^{-1/3}T^{4/5})|_{(75,15)}(-2) + 264S^{2/3}T^{-1/5}|_{(75,15)}(4.8) \\
&= 220(75)^{-1/3}(15)^{4/5}(-2) + 264(75)^{2/3}(15)^{-1/5}(4.8) \\
&= 12{,}201.4 \text{ units/month} \\
&\quad (\text{or } 508.392 \text{ units/day})
\end{aligned}$$

29. (a) First use the chain rule to find $\frac{\partial z}{\partial r}$ and $\frac{\partial z}{\partial \theta}$:

$$\begin{aligned}
\frac{\partial z}{\partial r} &= \frac{\partial z}{\partial x}\frac{\partial x}{\partial r} + \frac{\partial z}{\partial y}\frac{\partial y}{\partial r} \\
&= \frac{\partial z}{\partial x}(e^r \cos\theta) + \frac{\partial z}{\partial y}(e^r \sin\theta), \quad \text{and} \\
\frac{\partial z}{\partial \theta} &= \frac{\partial z}{\partial x}\frac{\partial x}{\partial \theta} + \frac{\partial z}{\partial y}\frac{\partial y}{\partial \theta} \\
&= \frac{\partial z}{\partial x}(-e^r \sin\theta) + \frac{\partial z}{\partial y}(e^r \cos\theta).
\end{aligned}$$

Now solve for $\frac{\partial z}{\partial x}$ and $\frac{\partial z}{\partial y}$:

$$\begin{aligned}
\frac{\partial z}{\partial x} &= e^{-r}\cos\theta\frac{\partial z}{\partial r} - e^{-r}\sin\theta\frac{\partial z}{\partial \theta}, \quad \text{and} \\
\frac{\partial z}{\partial y} &= e^{-r}\sin\theta\frac{\partial z}{\partial r} + e^{-r}\cos\theta\frac{\partial z}{\partial \theta}.
\end{aligned}$$

(b) Given the results for $\frac{\partial z}{\partial x}$ and $\frac{\partial z}{\partial y}$ in part (a), we compute:

$$\begin{aligned}
\frac{\partial^2 z}{\partial x^2} &= \frac{\partial}{\partial x}\left(\frac{\partial z}{\partial x}\right) = e^{-r}\cos\theta\frac{\partial}{\partial r}\left(\frac{\partial z}{\partial x}\right) - e^{-r}\sin\theta\frac{\partial}{\partial \theta}\left(\frac{\partial z}{\partial x}\right)\\
&= e^{-r}\cos\theta\frac{\partial}{\partial r}\left(e^{-r}\cos\theta\frac{\partial z}{\partial r} - e^{-r}\sin\theta\frac{\partial z}{\partial \theta}\right) - e^{-r}\sin\theta\frac{\partial}{\partial \theta}\left(e^{-r}\cos\theta\frac{\partial z}{\partial r} - e^{-r}\sin\theta\frac{\partial z}{\partial \theta}\right)\\
&= e^{-r}\cos\theta\left(-e^{-r}\cos\theta\frac{\partial z}{\partial r} + e^{-r}\cos\theta\frac{\partial^2 z}{\partial r^2} + e^{-r}\sin\theta\frac{\partial z}{\partial \theta} - e^{-r}\sin\theta\frac{\partial^2 z}{\partial r\partial\theta}\right)\\
&\quad - e^{-r}\sin\theta\left(-e^{-r}\sin\theta\frac{\partial z}{\partial r} + e^{-r}\cos\theta\frac{\partial^2 z}{\partial\theta\partial r} - e^{-r}\cos\theta\frac{\partial z}{\partial \theta} - e^{-r}\sin\theta\frac{\partial^2 z}{\partial \theta^2}\right)\\
&= e^{-2r}\left[(\sin^2\theta - \cos^2\theta)\frac{\partial z}{\partial r} + \cos^2\theta\frac{\partial^2 z}{\partial r^2} + 2\sin\theta\cos\theta\frac{\partial z}{\partial \theta} - 2\sin\theta\cos\theta\frac{\partial^2 z}{\partial r\partial\theta} + \sin^2\theta\frac{\partial^2 z}{\partial \theta^2}\right]
\end{aligned}$$

A similar calculation gives:

$$\begin{aligned}
\frac{\partial^2 z}{\partial y^2} &= \frac{\partial}{\partial y}\left(\frac{\partial z}{\partial y}\right) = e^{-r}\sin\theta\frac{\partial}{\partial r}\left(\frac{\partial z}{\partial y}\right) + e^{-r}\cos\theta\frac{\partial}{\partial \theta}\left(\frac{\partial z}{\partial y}\right)\\
&= e^{-r}\sin\theta\frac{\partial}{\partial r}\left(e^{-r}\sin\theta\frac{\partial z}{\partial r} + e^{-r}\cos\theta\frac{\partial z}{\partial \theta}\right) + e^{-r}\cos\theta\frac{\partial}{\partial \theta}\left(e^{-r}\sin\theta\frac{\partial z}{\partial r} + e^{-r}\cos\theta\frac{\partial z}{\partial \theta}\right)\\
&= e^{-2r}\left[(\cos^2\theta - \sin^2\theta)\frac{\partial z}{\partial r} + \sin^2\theta\frac{\partial^2 z}{\partial r^2} - 2\sin\theta\cos\theta\frac{\partial z}{\partial \theta} + 2\sin\theta\cos\theta\frac{\partial^2 z}{\partial r\partial\theta} + \cos^2\theta\frac{\partial^2 z}{\partial \theta^2}\right].
\end{aligned}$$

Now add these to get:

$$\frac{\partial^2 z}{\partial x^2} + \frac{\partial^2 z}{\partial y^2} = e^{-2r}[(\cos^2\theta + \sin^2\theta)z_{\theta\theta} + (\cos^2\theta + \sin^2\theta)z_{rr}] = e^{-2r}[z_{\theta\theta} + z_{rr}].$$

33. (a)

$$\begin{aligned}
\nabla^2(\nabla^2 f(x,y)) &= \frac{\partial^2}{\partial x^2}\left(\frac{\partial^2 f}{\partial x^2} + \frac{\partial^2 f}{\partial y^2}\right) + \frac{\partial^2}{\partial y^2}\left(\frac{\partial^2 f}{\partial x^2} + \frac{\partial^2 f}{\partial y^2}\right)\\
&= \frac{\partial^4 f}{\partial x^4} + \underbrace{\frac{\partial^4 f}{\partial x^2\partial y^2} + \frac{\partial^4 f}{\partial y^2\partial x^2}}_{\text{these are equal}-f\text{ is of class }C^4} + \frac{\partial^4 f}{\partial y^4}\\
&= \text{desired expression.}
\end{aligned}$$

(b) Similar:

$$\begin{aligned}
\nabla^2(\nabla^2 f) &= \frac{\partial^2}{\partial x_1^2}\left(\sum_{j=1}^n \frac{\partial^2 f}{\partial x_j^2}\right) + \cdots + \frac{\partial^2}{\partial x_n^2}\left(\sum_{j=1}^n \frac{\partial^2 f}{\partial x_j}\right)\\
&= \sum_{i=1}^n \frac{\partial^2}{\partial {x_i}^2}\left(\sum_{j=1}^n \frac{\partial^2 f}{\partial x_j^2}\right) = \sum_{i,j=1}^n \frac{\partial^4 f}{\partial x_i^2\partial x_j^2}.
\end{aligned}$$

35. $z = r\cos 3\theta$

(a) $z = r[\cos\theta\cos 2\theta - \sin\theta\sin 2\theta] = r[\cos\theta(\cos^2\theta - \sin^2\theta) - \sin\theta(2\sin\theta\cos\theta)]$ so

$$z = \frac{r^3[\cos^3\theta - \cos\theta\sin^2\theta - 2\sin^2\theta\cos\theta]}{r^2} = \frac{x^3 - 3xy^2}{x^2 + y^2}.$$

(b) Note that $\lim_{r\to 0} r\cos 3\theta = 0$ which is the value of the function at the origin. So yes, $f(x,y) = z$ is continuous at the origin.

(c) **(i)** $f_x = \dfrac{(x^2+y^2)(3x^2-3y^2)-(x^3-3xy^2)2x}{(x^2+y^2)^2} = \dfrac{x^4-3y^4+6x^2y^2}{(x^2+y^2)^2}$.

(ii) $f_y = \dfrac{(x^2+y^2)(-6xy)-(x^3-3xy^2)2y}{(x^2+y^2)^2} = \dfrac{-8x^3y}{(x^2+y^2)^2}$.

(iii) $f_x(0,0) = \lim_{h\to 0} \dfrac{f(h,0)-f(0,0)}{h} = \lim\limits_{h\to 0} \dfrac{h-0}{h} = 1$.

(iv) $f_y(0,0) = \lim_{h\to 0} \dfrac{f(0,h)-f(0,0)}{h} = \lim\limits_{h\to 0} \dfrac{0-0}{h} = 0$.

(d) $g(r,\theta) = r\cos 3\theta$ so $g_r(r,\theta) = \cos 3\theta$. This is the directional derivative $D_{\mathbf{u}}f$.

(e) When $(x,y) \neq (0,0)$, $f_y(x,y) = \dfrac{-8x^3y}{x^2+y^2}$. In particular, when $y = x$, $f_y = -2$. From part (c) $f_y(0,0) = 0$ so f_y is not continuous at the origin.

(f) Below are two sketches; the one on the left just shows a ribbon of the surface:

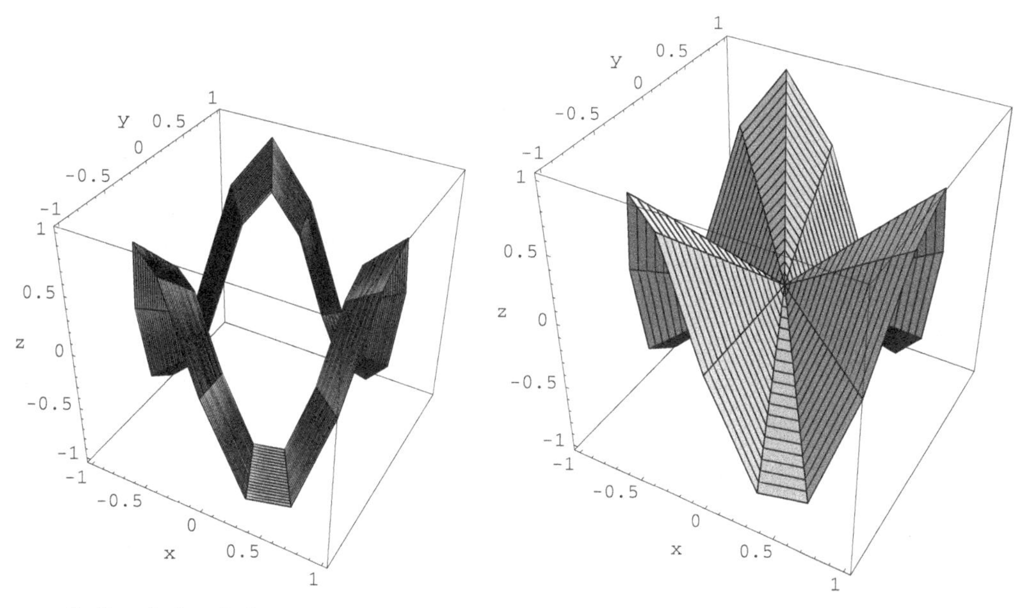

39. $F(tx,ty,tz) = t^3zy^2 - t^3x^3 + t^3x^2z = t^3F(x,y,z)$ so yes F is homogeneous of degree 3.

43. $F(tx_1,tx_2,\ldots,tx_n) = t^dF(x_1,x_2,\ldots,x_n)$ so that, by differentiating both sides with respect to t:

$$x_1\frac{\partial F}{\partial x_1}(tx_1,\ldots,tx_n) + \cdots + x_n\frac{\partial F}{\partial x_n}(tx_1,\ldots,tx_n) = dt^{d-1}F(x_1,\ldots,x_n).$$

Now let $t = 1$ and we get the result:

$$x_1\frac{\partial F}{\partial x_1} + \cdots + x_n\frac{\partial F}{\partial x_n} = dF.$$

Chapter 3

Vector-Valued Functions

3.1 Parametrized Curves and Kepler's Laws

3. This is the spiral $r = \theta$ (note $x = r\cos\theta$ and $y = r\sin\theta$):

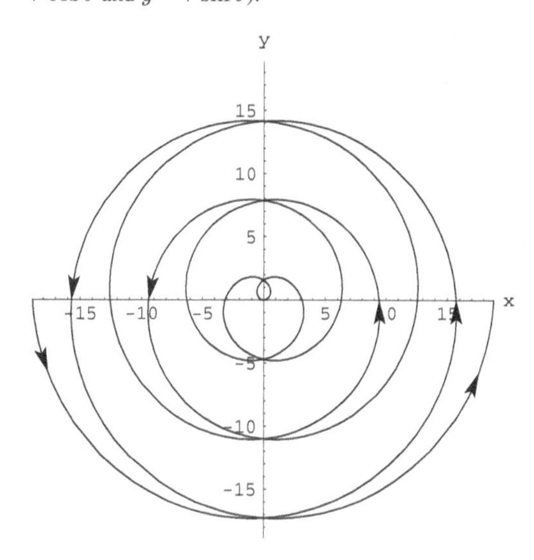

7. $\mathbf{x}(t) = (3t - 5, 2t + 7)$ so velocity $= \mathbf{v}(t) = \mathbf{x}'(t) = (3, 2)$ and speed $= \|\mathbf{v}(t)\| = \sqrt{3^2 + 2^2} = \sqrt{13}$. Finally, acceleration $= \mathbf{a}(t) = \mathbf{x}''(t) = (0, 0)$.

11. **(a)**

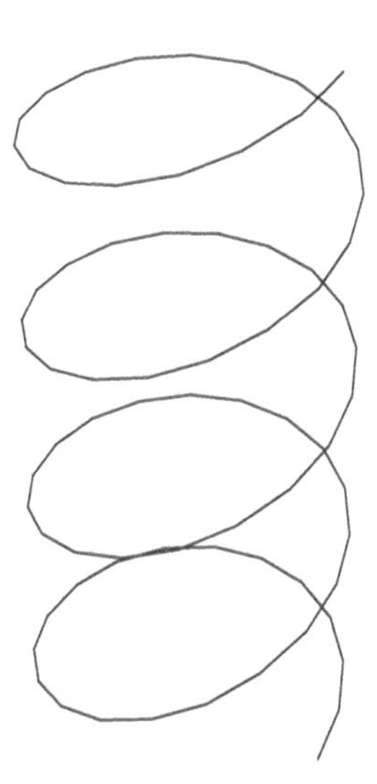

(b) To verify that the curve lies on the surface check that

$$\frac{x^2}{9} + \frac{y^2}{16} = \frac{9\cos^2 \pi t}{9} + \frac{16\sin^2 \pi t}{16} = \cos^2 \pi t + \sin^2 \pi t = 1.$$

The z component just determines the speed traveling up the cylinder.

15. Use formulas (2) and (3) from the text. $\mathbf{x}(t) = (te^{-t}, e^{3t})$ so $\mathbf{x}(0) = (0, 1)$ and $\mathbf{x}'(t) = (e^{-t} - te^{-t}, 3e^{3t})$ so $\mathbf{x}'(0) = (1, 3)$. The equation of the tangent line at $t = 0$ is $\mathbf{l}(t) = (0, 1) + (1, 3)t = (t, 1 + 3t)$.

19. **(a)** The sketch of $\mathbf{x}(t) = (t, t^3 - 2t + 1)$ is:

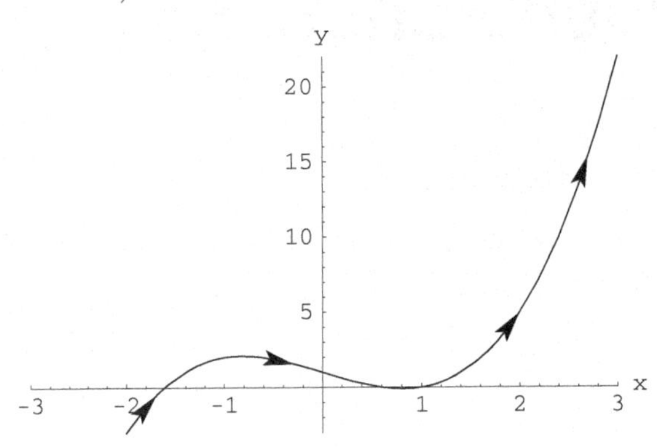

(b) $\mathbf{x}(2) = (2, 5)$ and since $\mathbf{x}'(t) = (1, 3t^2 - 2)$ we get $\mathbf{x}'(2) = (1, 10)$. The equation of the line is then

$$\mathbf{l}(t) = (2, 5) + (1, 10)(t - 2) = (t, 10t - 15).$$

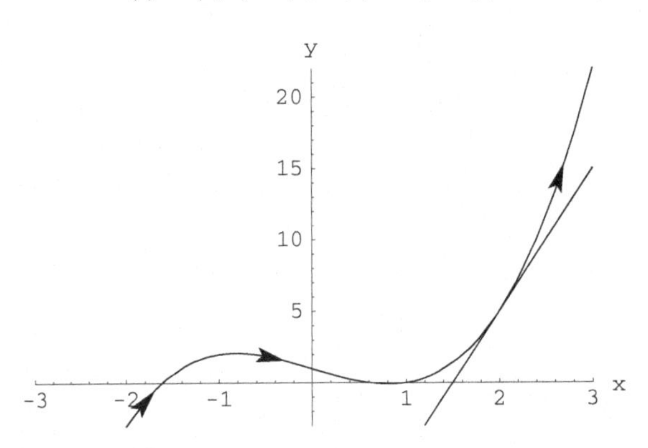

(c) Since $x = t$ we see that $y = f(x) = x^3 - 2x + 1$.

(d) So the equation of the tangent line at $x = 2$ is $y - f(2) = f'(2)(x - 2)$, where f is as in part (c). Substituting, we get $y - 5 = 10(x - 2)$ or $y = 10x - 15$. This is consistent with our answer for part (b).

23. We use formula (5) and solve for θ. So, $x = v_0^2 \sin 2\theta / g$ becomes $1500 = 250^2 \sin 2\theta/32$ or

$$\sin 2\theta = \frac{(1500)(32)}{62500} = \frac{96}{125}.$$

There are two values of θ with $0 \leq \theta \leq \pi/2$ that satisfy this last equation. One is

$$\theta = (1/2)\sin^{-1}(96/125) \approx 0.43786 \approx 25.088°,$$

and the other is

$$\theta = \pi/2 - (1/2)\sin^{-1}(96/125) \approx 1.13294 \approx 64.913°.$$

25. We have $\mathbf{x}(2) = \left(e^4, 8, \frac{3}{2}\right)$ and $\mathbf{x}'(2) = \left(2e^4, 10, \frac{5}{4}\right)$. If the rocket's engines cease when $t = 2$, then the rocket will follow the tangent line path

$$\mathbf{l}(t) = \mathbf{x}(2) + (t - 2)\mathbf{x}'(2) = \left(e^4(2t - 3), 10t - 12, \tfrac{5}{4}t - 1\right).$$

For this path to reach the space station, we must have

$$\left(e^4(2t - 3), 10t - 12, \frac{5}{4}t - 1\right) = (7e^4, 35, 5).$$

Thus, in particular

$$e^4(2t - 3) = 7e^4 \Leftrightarrow 2t - 3 = 7 \Leftrightarrow t = 5.$$

However $\mathbf{l}(5) = \left(7e^4, 38, \frac{21}{4}\right) \neq (7e^4, 35, 5)$. Hence the rocket does *not* reach the repair station.

29. You're asked to show that if $\|\mathbf{x}(t)\|$ is constant, then $\mathbf{x}$ is perpendicular to $d\mathbf{x}/dt$. If $\|\mathbf{x}(t)\|$ is constant, then $\frac{d}{dt}\|\mathbf{x}(t)\| \equiv 0$. So

$$0 = \frac{d}{dt}\|\mathbf{x}(t)\| = \frac{d}{dt}\sqrt{\mathbf{x}\cdot\mathbf{x}} = \left(\frac{1}{2\sqrt{\mathbf{x}\cdot\mathbf{x}}}\right)\left(2\frac{d\mathbf{x}}{dt}\cdot\mathbf{x}\right).$$

This means that $\frac{d\mathbf{x}}{dt}\cdot\mathbf{x} = 0$.

33. **(a)** To have $\mathbf{x}(t_1) = (t_1^2, t_1^3 - t_1) = (t_2^2, t_2^3 - t_2) = \mathbf{x}(t_2)$, we must have $t_1^2 = t_2^2$, so if $t_1 \neq t_2$, then $t_1 = -t_2$. Then, comparing the second components: $t_1^3 - t_1 = -t_1^3 + t_1 \iff 2t_1^3 = 2t_1$. Since $t_1 \neq 0$ (otherwise $t_2 = 0$ as well), we must have $t_1^2 = 1$. Thus $\mathbf{x}(1) = \mathbf{x}(-1) = (1, 0)$.

(b) The velocity vector of the path is $\mathbf{x}'(t) = (2t, 3t^2 - 1)$. Therefore, the corresponding tangent vectors at $t = \pm 1$ are $\mathbf{x}'(-1) = (-2, 2)$ and $\mathbf{x}'(1) = (2, 2)$. Note that $\mathbf{x}'(-1)\cdot\mathbf{x}'(1) = 0$. Since these tangent vectors are parallel to the corresponding tangent lines, we see that the tangent lines must be perpendicular—so the angle they make is $\pi/2$.

3.2 Arclength and Differential Geometry

3. $\mathbf{x}(t) = (\cos 3t, \sin 3t, 2t^{3/2})$ so $\mathbf{x}'(t) = (-3\sin 3t, 3\cos 3t, 3t^{1/2})$. The length of the path is then

$$L(\mathbf{x}) = \int_0^2 \sqrt{9\sin^2 3t + 9\cos^2 3t + 9t}\,dt = 3\int_0^2 \sqrt{1+t}\,dt = 3\int_1^3 \sqrt{u}\,du = 2u^{3/2}\big|_1^3 = 6\sqrt{3} - 2.$$

5. $\mathbf{x}(t) = (t^3, 3t^2, 6t)$ so $\mathbf{x}'(t) = (3t^2, 6t, 6)$. The length of the path is then

$$L(\mathbf{x}) = \int_{-1}^2 \sqrt{9t^4 + 36t^2 + 36}\,dt = \int_{-1}^2 \sqrt{9(t^2+2)^2}\,dt$$
$$= \int_{-1}^2 3(t^2+2)\,dt = (t^3 + 6t)\Big|_{-1}^2 = 27.$$

9. A sketch of the curve $\mathbf{x}(t) = (a\cos^3 t, a\sin^3 t)$ for $0 \le t \le 2\pi$ is:

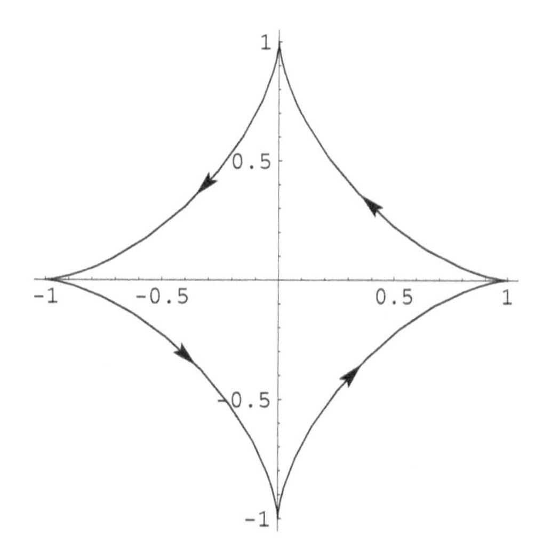

Because of the obvious symmetries we will compute the length of the portion of the curve in the first quadrant and multiply it by 4:

$$L(\mathbf{x}) = 4\int_0^{\pi/2} \|(-3a\cos^2 t\sin t, 3a\sin^2 t\cos t)\|\,dt = 4\int_0^{\pi/2} \sqrt{9a^2(\cos^4 t\sin^2 t + \sin^4 t\cos^2 t)}\,dt$$
$$= 4\int_0^{\pi/2} \sqrt{9a^2\sin^2 t\cos^2 t(\cos^2 t + \sin^2 t)}\,dt = 4\int_0^{\pi/2} 3a\sin t\cos t\,dt = 6a\sin^2 t\big|_0^{\pi/2}$$
$$= 6a.$$

13. **(a)** A sketch of $\mathbf{x} = |t-1|\mathbf{i} + |t|\mathbf{j}, -2 \le t \le 2$ is:

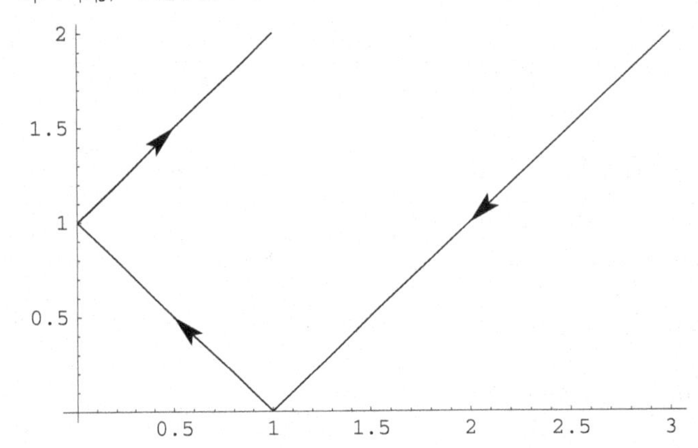

(b) Except for two points, the path is smooth (more than C^1). In fact, the path is comprised of three line segments joined end to end. In other words, on the open intervals $-2 \le t < 0, 0 < t < 1$, and $1 < t \le 2$, the path $\mathbf{x}$ is C^1. We say that the path $\mathbf{x}$ is piecewise C^1.

(c) We could figure out the length of each piece and add them together. In the process we will find that we're working too hard.

$$\mathbf{x}(t) = \begin{cases} (1-t)\mathbf{i} - t\mathbf{j} & -2 \le t \le 0 \\ (1-t)\mathbf{i} + t\mathbf{j} & 0 < t \le 1 \\ (t-1)\mathbf{i} + t\mathbf{j} & 1 < t \le 2 \end{cases} \quad \text{so} \quad \mathbf{x}'(t) = \begin{cases} (-1,-1) & -2 \le t \le 0 \\ (-1,1) & 0 < t \le 1 \\ (1,1) & 1 < t \le 2 \end{cases}.$$

So we see that $\|\mathbf{x}'(t)\| \equiv \sqrt{2}$. This means that to calculate the length of the curve we don't have to break up the integral into three pieces:

$$L(\mathbf{x}) = \int_{-2}^{2} \sqrt{2}\, dt = 4\sqrt{2}.$$

19. $\mathbf{x}(t) = (t, (1/3)(t+1)^{3/2}, (1/3)(1-t)^{3/2})$ so $\mathbf{x}'(t) = (1, (1/2)(t+1)^{1/2}, -(1/2)(1-t)^{1/2})$, and $\|\mathbf{x}'(t)\| = \sqrt{3/2}$.

$$\mathbf{T} = \sqrt{\frac{2}{3}}\left(1, \tfrac{1}{2}\sqrt{t+1}, -\tfrac{1}{2}\sqrt{1-t}\right), \text{ and}$$

$$\mathbf{N} = \frac{\sqrt{2/3}(0, (1/4)(t+1)^{-1/2}, (1/4)(1-t)^{-1/2})}{\sqrt{(2/3)(1/16)\left(\frac{1}{t+1} + \frac{1}{1-t}\right)}}$$

$$= \frac{1}{\sqrt{2}}(0, \sqrt{1-t}, \sqrt{t+1}), \text{ and}$$

$$\mathbf{B} = \sqrt{\frac{1}{3}}\left(\tfrac{1}{2}(t+1) + \tfrac{1}{2}(1-t), -\sqrt{t+1}, \sqrt{1-t}\right) = \frac{1}{\sqrt{3}}(1, -\sqrt{t+1}, \sqrt{1-t}).$$

Also,

$$\kappa = \frac{\|\sqrt{2/3}(0, (1/4)(t+1)^{-1/2}, (1/4)(1-t)^{-1/2})\|}{\sqrt{3/2}} = \frac{1}{3\sqrt{2(1-t^2)}}.$$

Finally,

$$\frac{d\mathbf{B}}{dt} = \frac{1}{\sqrt{3}}\left(0, -\frac{1}{2\sqrt{t+1}}, -\frac{1}{2\sqrt{1-t}}\right) \text{ so}$$

$$\frac{d\mathbf{B}}{ds} = \frac{1}{\sqrt{3}}\left(0, -\frac{1}{2\sqrt{t+1}}, -\frac{1}{2\sqrt{1-t}}\right) \Big/ \sqrt{3/2} = -\tau\mathbf{N}.$$

Solving,

$$\tau = \frac{1}{3\sqrt{(1-t^2)}}.$$

21. **(a)** By formula (17): $\kappa = \frac{\|\mathbf{x}' \times \mathbf{x}''\|}{\|\mathbf{x}'\|^3}$. Let $y = f(x)$ and view the problem as sitting inside of $\mathbf{R}^3$. Then $\mathbf{x} = (x, f(x), 0), \mathbf{x}' = (1, f'(x), 0)$, and $\mathbf{x}'' = (0, f''(x), 0)$. We calculate the cross product $\mathbf{x}' \times \mathbf{x}'' = (0, 0, f''(x))$ so

$$\kappa = \frac{\|(0, 0, f''(x))\|}{\|(1, f'(x), 0)\|^3} = \frac{|f''(x)|}{[1 + (f'(x))^2]^{3/2}}.$$

(b) If $y = \ln(\sin x)$, then $y' = \cos x / \sin x$ and $y'' = -1/\sin^2 x$. By our results for part (a),

$$\kappa = \frac{|-1/\sin^2 x|}{[1 + (\cos^2 x/\sin^2 x)]^{3/2}} = |\sin x|.$$

25. **(a)** The curvature is calculated to be (with some simplification)

$$\frac{3(1 + \cos t)}{16 \cos^3(t/2)}.$$

(b) The path is pictured below left while the corresponding curvature is plotted below right.

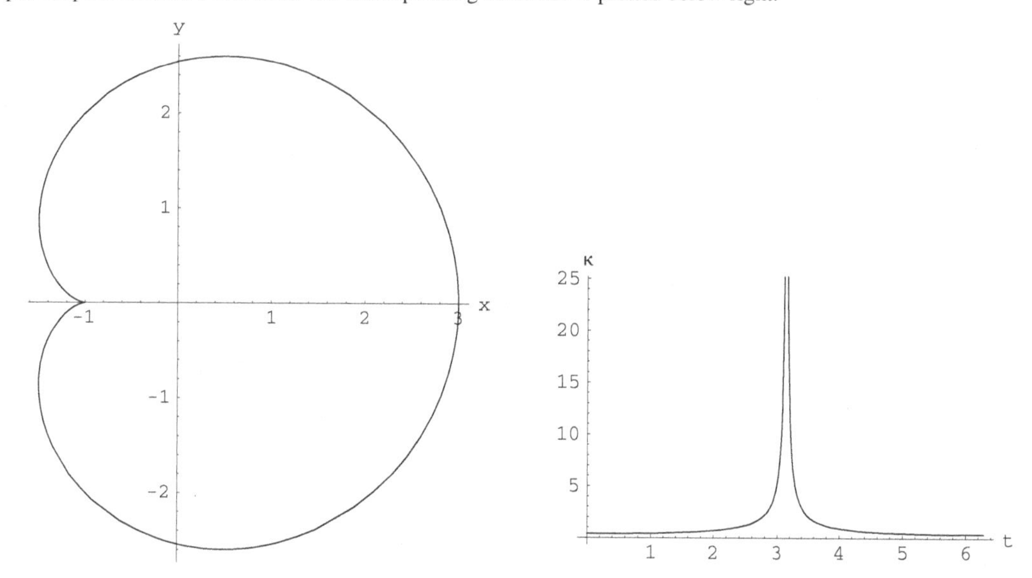

27. $\mathbf{x}(t) = (t^2, t)$ so $\mathbf{x}'(t) = (2t, 1)$ and $\mathbf{x}''(t) = (2, 0)$. The speed is then $\|\mathbf{x}'(t)\| = \sqrt{1 + 4t^2}$ and so the tangential component of acceleration is $\ddot{s} = 4t/\sqrt{1 + 4t^2}$. Since $\|\mathbf{a}\| = 2, \|\mathbf{a}\|^2 - \ddot{s}^2 = 4/(1 + 4t^2)$, so the normal component of acceleration is $2/\sqrt{1 + 4t^2}$.

31. $\mathbf{x}(t) = (t, t, t^2)$ so $\mathbf{x}'(t) = (1, 1, 2t)$ and $\mathbf{x}''(t) = (0, 0, 2)$. The speed is then $\|\mathbf{x}'(t)\| = \sqrt{2 + 4t^2}$ and so the tangential component of acceleration is $\ddot{s} = 4t/\sqrt{2 + 4t^2}$. Since $\|\mathbf{a}\| = 2, \|\mathbf{a}\|^2 - \ddot{s}^2 = 4/(1 + 2t^2)$, so the normal component of acceleration is $2/\sqrt{1 + 2t^2}$.

35. To establish the formula, first note that $\mathbf{v} \times \mathbf{a} = \kappa \dot{s}^3 \mathbf{B}$ (see, for example, the calculation leading up to formula (17)) and $\|\mathbf{v} \times \mathbf{a}\| = \kappa \dot{s}^3 = \kappa \|\mathbf{v}\|^3$. So

$$\frac{(\mathbf{v} \times \mathbf{a}) \cdot \mathbf{a}'}{\|\mathbf{v} \times \mathbf{a}\|^2} = \frac{\kappa \dot{s}^3 \mathbf{B} \cdot \mathbf{a}'}{\kappa^2 \dot{s}^6} = \frac{\mathbf{B} \cdot \mathbf{a}'}{\kappa \dot{s}^3}.$$

Now, $\mathbf{a}(t) = \ddot{s}\mathbf{T} + \kappa \dot{s}^2 \mathbf{N}$ and by the Frenet equations $\mathbf{N}'(s) = -\kappa \mathbf{T} + \tau \mathbf{B}$. Since we are calculating the dot product of $\mathbf{a}'$ with $\mathbf{B}$, the only piece that will survive is the coefficient of $\mathbf{B}$, so

$$\begin{aligned} \mathbf{a}'(t) &= (\text{something without } \mathbf{B}) + \kappa \dot{s}^2 \mathbf{N}'(s) \frac{ds}{dt} \\ &= (\text{something else without } \mathbf{B}) + \kappa \dot{s}^3 \tau \mathbf{B} \end{aligned}$$

and so, putting it all together,

$$\frac{(\mathbf{v} \times \mathbf{a}) \cdot \mathbf{a}'}{\|\mathbf{v} \times \mathbf{a}\|^2} = \frac{\kappa \dot{s}^3 \mathbf{B} \cdot \mathbf{a}'}{\kappa^2 \dot{s}^6} = \frac{\mathbf{B} \cdot \mathbf{a}'}{\kappa \dot{s}^3} = \frac{\mathbf{B} \cdot \kappa \dot{s}^3 \tau \mathbf{B}}{\kappa \dot{s}^3} = \tau.$$

39. We have $\|\mathbf{x}-\mathbf{x}_0\|^2 = (\mathbf{x}-\mathbf{x}_0)\cdot(\mathbf{x}-\mathbf{x}_0) = a^2$. Thus $\|\mathbf{x}-\mathbf{x}_0\| = a$, so $\mathbf{x}(t)$ lies on a sphere of radius a.

43.

$$\|\mathbf{w}\|^2 = \mathbf{w}\cdot\mathbf{w} = (\tau\mathbf{T}+\kappa\mathbf{B})\cdot(\tau\mathbf{T}+\kappa\mathbf{B}) = \tau^2\mathbf{T}\cdot\mathbf{T} + \kappa\tau\mathbf{B}\cdot\mathbf{T} + \kappa\tau\mathbf{T}\cdot\mathbf{B} + \kappa^2\mathbf{B}\cdot\mathbf{B} = \tau^2+\kappa^2$$

3.3 Vector Fields: An Introduction

1. $\mathbf{F} = y\mathbf{i} - x\mathbf{j} = (y, -x)$ is shown below. The figure can be generated using *Mathematica* or Maple. The axes are the 'usual' positions with the origin at the center. The relative length of the shart of the arrows corresponds to the length of the vectors.

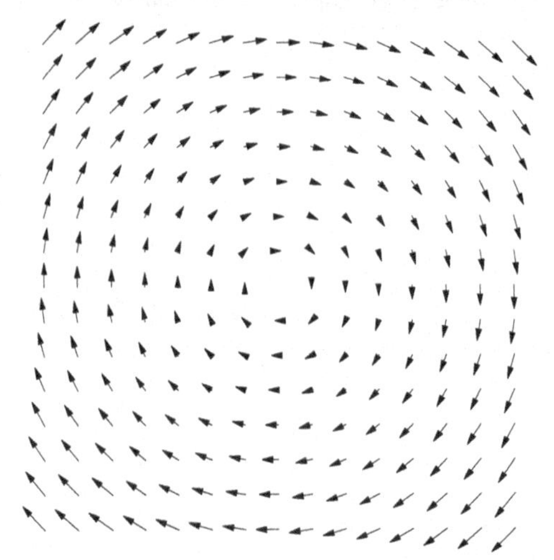

3. $\mathbf{F} = (-x, y)$ is shown below.

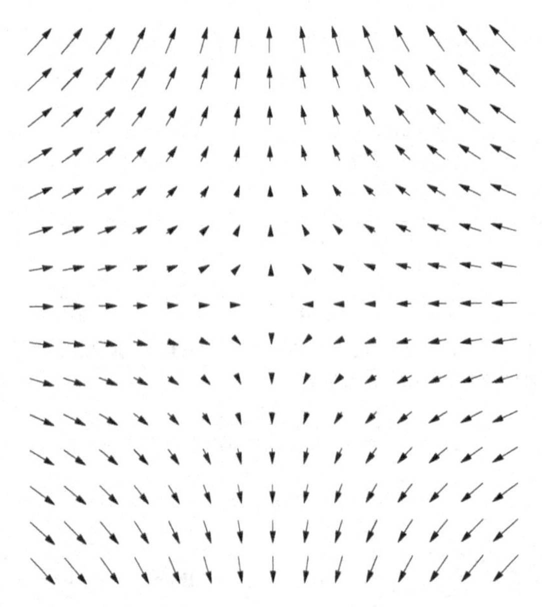

7. $\mathbf{F} = 3\mathbf{i} + 2\mathbf{j} + \mathbf{k} = (3, 2, 1)$ is constant. The figure on the right shows the slice in the xy-plane:

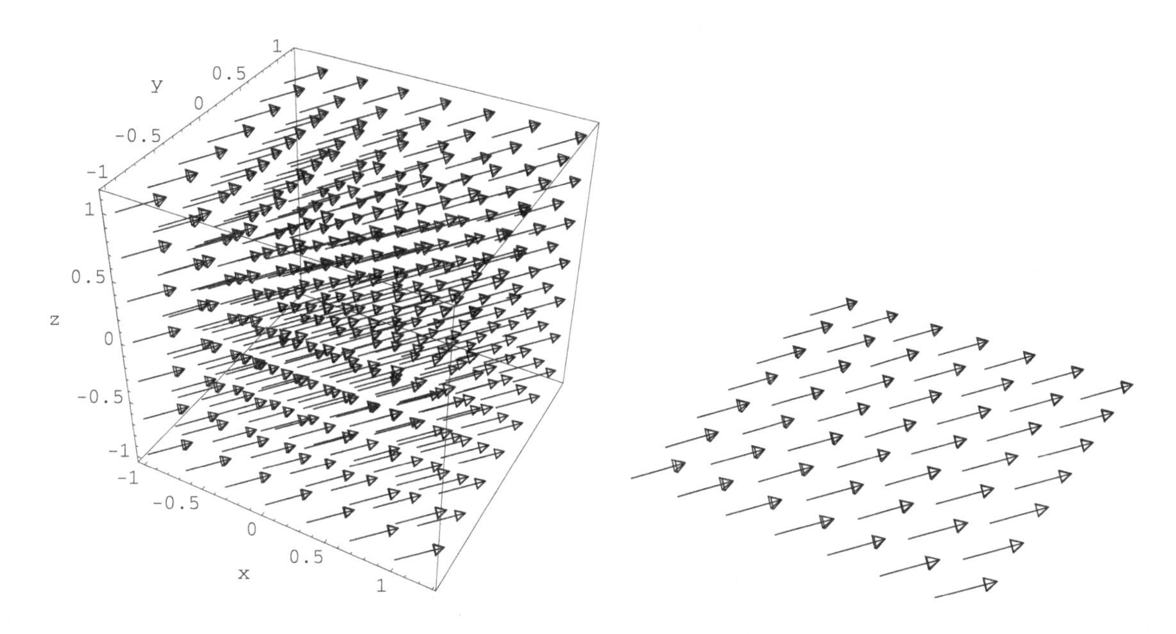

11. $\mathbf{F} = (y, -x, z)$. The figure on the right shows the slices in the $z = 1$ and $z = -1$ planes—compare this to Exercises 8 and 10:

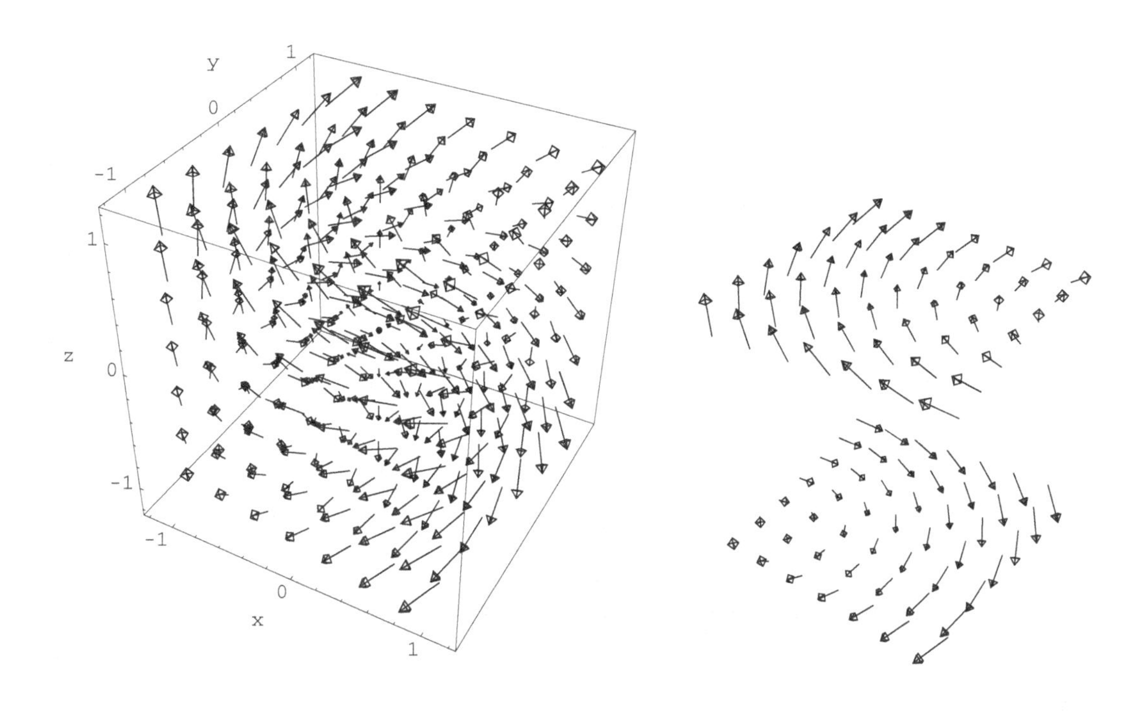

15. The figure is below.

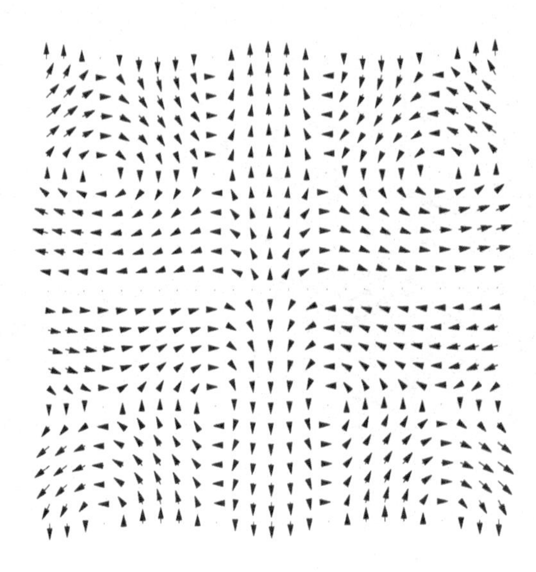

17. $\mathbf{x}(t) = (x, y, z) = (\sin t, \cos t, 0)$ so $\mathbf{x}'(t) = (\cos t, -\sin t, 0) = (y, -x, 0) = \mathbf{F}(\mathbf{x}(t))$. We can see below how the path, in bold, is a flow line for the vector field we saw above in Exercise 8. The figure on the right is the xy-plane slice of the figure on the left.

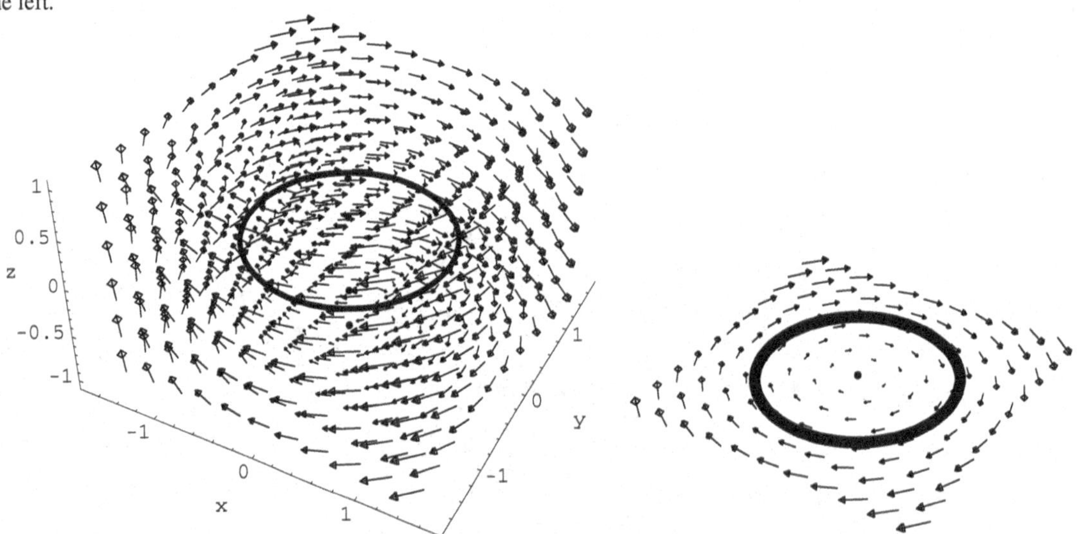

21. If $\mathbf{x}(t) = (x, y)$ then $\mathbf{x}'(t) = \mathbf{F}(x, y) = (x^2, y)$. As in Exercise 20, we know that $y = ke^t$ and $y(1) = e$ tells us that $k = 1$. As for x, $dx/dt = x^2$. This is a separable differential equation $dx/x^2 = dt$. Integrating and solving for x gives us $x = -1/(t + c)$. From the initial condition $x(1) = 1$ we find that $1 = -1/(1 + c)$ or $c = -2$. The equation, therefore, of the flow line is $\mathbf{x}(t) = (1/(2 - t), e^t)$.

23. **(a)** For the function $f(x, y, z) = 3x - 2y + z$, $\nabla f = \mathbf{F}$ so $\mathbf{F}$ is a gradient field.

(b) The equipotiential surfaces are those for which $f(x, y, z)$ is constant. $3x - 2y + z = c$. These are planes with normal vector $(3, -2, 1)$.

27. First we see that $\phi(x, y, 0) = (y \sin 0 + x \cos 0, y \cos 0 - x \sin 0) = (x, y)$. Next,

$$\begin{aligned}\frac{\partial}{\partial t}\phi(x, y, t) &= (y \cos t - x \sin t, -y \sin t - x \cos t) \\ &= \phi(y, -x, t) = \mathbf{F}(\phi(x, y, t)).\end{aligned}$$

31. We know that $\frac{\partial}{\partial t}\phi(\mathbf{x},t) = \mathbf{F}(\phi(\mathbf{x},t))$. So

$$\frac{\partial}{\partial t} D_{\mathbf{x}}\phi(\mathbf{x},t) = D_{\mathbf{x}}\left(\frac{\partial}{\partial t}\phi(\mathbf{x},t)\right) = D_{\mathbf{x}}\mathbf{F}(\phi(\mathbf{x},t)).$$

Now by the chain rule (Theorem 5.3):

$$D_{\mathbf{x}}\mathbf{F}(\phi(\mathbf{x},t)) = D\mathbf{F}(\phi(\mathbf{x},t))D_{\mathbf{x}}\phi(\mathbf{x},t).$$

3.4 Gradient, Divergence, Curl, and The Del Operator

3. $\mathbf{F} = (x+y, y+z, x+z)$, so $\operatorname{div}\mathbf{F} = \nabla\cdot\mathbf{F} = \frac{\partial F_1}{\partial x} + \frac{\partial F_2}{\partial y} + \frac{\partial F_3}{\partial z} = 1+1+1 = 3$.

5. $\mathbf{F} = (x_1^2, 2x_2^2, \ldots, nx_n^2)$, so $\operatorname{div}\mathbf{F} = \nabla\cdot\mathbf{F} = \frac{\partial F_1}{\partial x_1} + \frac{\partial F_2}{\partial x_2} + \cdots + \frac{\partial F_n}{\partial x_n} = 2x_1 + 4x_2 + \cdots + 2nx_n$.

9. $\operatorname{curl}\mathbf{F} = \begin{vmatrix} \mathbf{i} & \mathbf{j} & \mathbf{k} \\ \frac{\partial}{\partial x} & \frac{\partial}{\partial y} & \frac{\partial}{\partial z} \\ x+yz & y+xz & z+xy \end{vmatrix} = (x-x, -y+y, z-z) = (0,0,0)$.

13. **(a)** At each point "more is moving away than towards" so div $\mathbf{F} > 0$ on all $\mathbf{R}^2$.

(b) At each point "more is moving towards than away" so div $\mathbf{F} < 0$ on all $\mathbf{R}^2$.

(c) Here we have a mixed bag. At each point to the left of the y-axis "more is moving towards than away" and at each point to the right of the y-axis "more is moving away than towards" so div $\mathbf{F} < 0$ for $x < 0$, div $\mathbf{F} > 0$ for $x > 0$, and div $\mathbf{F} = 0$ for $x = 0$.

(d) Again we have a mixed bag. At each point the above the x-axis "more is moving towards than away" and at each point below the x-axis "more is moving away than towards" so div $\mathbf{F} < 0$ for $y > 0$, div $\mathbf{F} > 0$ for $y < 0$, and div $\mathbf{F} = 0$ for $y = 0$.

15. $\mathbf{F}(x,y,z) = xyz\mathbf{i} + e^z\cos x\mathbf{j} + xy^2z^3\mathbf{k}$ so

$$\nabla\times\mathbf{F} = \begin{vmatrix} \mathbf{i} & \mathbf{j} & \mathbf{k} \\ \frac{\partial}{\partial x} & \frac{\partial}{\partial y} & \frac{\partial}{\partial z} \\ xyz & e^z\cos x & xy^2z^3 \end{vmatrix} = (2xyz^3 + e^z\cos x, -y^2z^3 + xy, e^z\sin x - xz).$$

Finally we calculate

$$\begin{aligned}\nabla\cdot(\nabla\times\mathbf{F}) &= \frac{\partial}{\partial x}(2xyz^3 + e^z\cos x) + \frac{\partial}{\partial y}(-y^2z^3 + xy) + \frac{\partial}{\partial z}(e^z\sin x - xz) \\ &= 2yz^3 - e^z\sin x - 2yz^3 + x + e^z\sin x - x = 0.\end{aligned}$$

19. In this exercise we'll need to know that $r^n\mathbf{r} = (x^2+y^2+z^2)^{n/2}(x,y,z)$.

$$\begin{aligned}\nabla\cdot(r^n\mathbf{r}) &= \frac{\partial}{\partial x}[x(x^2+y^2+z^2)^{n/2}] + \frac{\partial}{\partial y}[y(x^2+y^2+z^2)^{n/2}] + \frac{\partial}{\partial z}[z(x^2+y^2+z^2)^{n/2}] \\ &= \left[r^n + x\left(\frac{n}{2}\right)2x(x^2+y^2+z^2)^{(n-2)/2}\right] + \left[r^n + y\left(\frac{n}{2}\right)2y(x^2+y^2+z^2)^{(n-2)/2}\right] \\ &\quad + \left[r^n + z\left(\frac{n}{2}\right)2z(x^2+y^2+z^2)^{(n-2)/2}\right] \\ &= [r^n + nx^2r^{n-2}] + [r^n + ny^2r^{n-2}] + [r^n + nz^2r^{n-2}] \\ &= 3r^n + n(x^2+y^2+z^2)r^{n-2} = 3r^n + nr^2r^{n-2} = 3r^n + nr^n = (n+3)r^n\end{aligned}$$

23.

$$\begin{aligned}\nabla\cdot(f\mathbf{F}) &= \sum_{i=1}^{n}\frac{\partial}{\partial x_i}(fF_i) = \sum_{i=1}^{n}\left(\frac{\partial f}{\partial x_i}F_i + f\frac{\partial F_i}{\partial x_i}\right) \\ &= \sum_{i=1}^{n}\left(\frac{\partial f}{\partial x_i}F_i\right) + \sum_{i=1}^{n}\left(f\frac{\partial F_i}{\partial x_i}\right) = \nabla(f)\cdot\mathbf{F} + f(\nabla\cdot\mathbf{F}) \\ &= f\nabla\cdot\mathbf{F} + \mathbf{F}\cdot\nabla f.\end{aligned}$$

27. We will need formula (9) from Section 1.7:

$$\begin{cases} \mathbf{e}_\rho = \sin\varphi\cos\theta\mathbf{i} + \sin\varphi\sin\theta\mathbf{j} + \cos\varphi\mathbf{k} \\ \mathbf{e}_\varphi = \cos\varphi\cos\theta\mathbf{i} + \cos\varphi\sin\theta\mathbf{j} - \sin\varphi\mathbf{k} \\ \mathbf{e}_\theta = -\sin\theta\mathbf{i} + \cos\theta\mathbf{j}. \end{cases}$$

From the chain rule, we have the following relations between rectangular and spherical differential operators:

$$\begin{cases} \dfrac{\partial}{\partial\rho} = \sin\varphi\cos\theta\dfrac{\partial}{\partial x} + \sin\varphi\sin\theta\dfrac{\partial}{\partial y} + \cos\varphi\dfrac{\partial}{\partial z} \\ \dfrac{\partial}{\partial\varphi} = \rho\cos\varphi\cos\theta\dfrac{\partial}{\partial x} + \rho\cos\varphi\sin\theta\dfrac{\partial}{\partial y} - \rho\sin\varphi\dfrac{\partial}{\partial z} \\ \dfrac{\partial}{\partial\theta} = -\rho\sin\varphi\sin\theta\dfrac{\partial}{\partial x} + \rho\sin\varphi\cos\theta\dfrac{\partial}{\partial y}. \end{cases}$$

Solving for $\partial/\partial x, \partial/\partial y$, and $\partial/\partial z$:

$$\begin{cases} \dfrac{\partial}{\partial x} = \sin\varphi\cos\theta\dfrac{\partial}{\partial\rho} + \dfrac{\cos\varphi\cos\theta}{\rho}\dfrac{\partial}{\partial\varphi} - \dfrac{\sin\theta}{\rho\sin\varphi}\dfrac{\partial}{\partial\theta} \\ \dfrac{\partial}{\partial y} = \sin\varphi\sin\theta\dfrac{\partial}{\partial\rho} + \dfrac{\cos\varphi\sin\theta}{\rho}\dfrac{\partial}{\partial\varphi} + \dfrac{\cos\theta}{\rho\sin\varphi}\dfrac{\partial}{\partial\theta} \\ \dfrac{\partial}{\partial z} = \cos\varphi\dfrac{\partial}{\partial\rho} - \dfrac{\sin\varphi}{\rho}\dfrac{\partial}{\partial\varphi}. \end{cases}$$

Now we calculate the gradient:

$$\begin{aligned} \nabla f &= \frac{\partial f}{\partial x}\mathbf{i} + \frac{\partial f}{\partial y}\mathbf{j} + \frac{\partial f}{\partial z}\mathbf{k} \\ &= \left(\sin\varphi\cos\theta\frac{\partial f}{\partial\rho} + \frac{\cos\varphi\cos\theta}{\rho}\frac{\partial f}{\partial\varphi} - \frac{\sin\theta}{\rho\sin\varphi}\frac{\partial f}{\partial\theta}\right)\mathbf{i} \\ &\quad + \left(\sin\varphi\sin\theta\frac{\partial f}{\partial\rho} + \frac{\cos\varphi\sin\theta}{\rho}\frac{\partial f}{\partial\varphi} + \frac{\cos\theta}{\rho\sin\varphi}\frac{\partial f}{\partial\theta}\right)\mathbf{j} + \left(\cos\varphi\frac{\partial f}{\partial\rho} - \frac{\sin\varphi}{\rho}\frac{\partial f}{\partial\varphi}\right)\mathbf{k} \\ &= \frac{\partial f}{\partial\rho}(\sin\varphi\cos\theta\mathbf{i} + \sin\varphi\sin\theta\mathbf{j} + \cos\varphi\mathbf{k}) + \left(\frac{1}{\rho}\right)\frac{\partial f}{\partial\varphi}(\cos\varphi\cos\theta\mathbf{i} + \cos\varphi\sin\theta\mathbf{j} - \sin\varphi\mathbf{k}) \\ &\quad + \left(\frac{1}{\rho\sin\varphi}\right)\frac{\partial f}{\partial\theta}(-\sin\theta\mathbf{i} + \cos\theta\mathbf{j}) = \frac{\partial f}{\partial\rho}\mathbf{e}_r + \left(\frac{1}{\rho}\right)\frac{\partial f}{\partial\varphi}\mathbf{e}_\varphi + \left(\frac{1}{\rho\sin\varphi}\right)\frac{\partial f}{\partial\theta}\mathbf{e}_\theta. \end{aligned}$$

31. (a) Let $\mathbf{G}(t) = \mathbf{F}(\mathbf{a} + t\mathbf{v})$. Then

$$\begin{aligned} D_\mathbf{v}\mathbf{F}(\mathbf{a}) = \lim_{t\to 0}\frac{1}{t}(\mathbf{F}(\mathbf{a} + t\mathbf{v}) - \mathbf{F}(\mathbf{a})) &= \lim_{t\to 0}\frac{1}{t}(\mathbf{G}(t) - \mathbf{G}(0)) \\ &= \mathbf{G}'(0). \end{aligned}$$

Thus

$$D_\mathbf{v}\mathbf{F}(\mathbf{a}) = \left.\frac{d}{dt}\mathbf{F}(\mathbf{a} + t\mathbf{v})\right|_{t=0}.$$

(b) $\frac{d}{dt}\mathbf{F}(\mathbf{a} + t\mathbf{v}) = D\mathbf{F}(\mathbf{a} + t\mathbf{v})\frac{d}{dt}(\mathbf{a} + t\mathbf{v}) = D\mathbf{F}(\mathbf{a} + t\mathbf{v})\mathbf{v}$. Now evaluate at $t = 0$ to get $D_\mathbf{v}\mathbf{F}(\mathbf{a}) = D\mathbf{F}(\mathbf{a})\mathbf{v}$.

True/False Exercises for Chapter 3

1. True.
3. True.
5. False. (There should be a negative sign in the second term on the right.)
7. True.
9. False.

11. True.

13. True.

15. False. (It's a scalar field.)

17. True.

19. False. (It's a meaningless expression.)

21. True. (Check that $\mathbf{F}(\mathbf{x}(t)) = \mathbf{x}'(t)$.)

23. False. ($\nabla \times \mathbf{F} \neq \mathbf{0}$.)

25. False. (Consider $\mathbf{F} = y\mathbf{i} + x\mathbf{j}$.)

27. True.

29. False. ($\nabla \cdot (\nabla \times \mathbf{F}) \neq 0$.)

Miscellaneous Exercises for Chapter 3

1. Here are the answers: (a) D (b) F (c) A
(d) B (e) C (f) E

Here's some explanation: The formulas in (a) and in (f) are the only ones that keep x and y bounded (between -1 and 1), so they must correspond to D and E. Note that in (a) $\mathbf{x}(0) = (0, 0)$, but the graph in E does not pass through the origin. Note that in (c) $x \geq 1$ and the only graph with that property is A. In (b) we see that $\mathbf{x}(-t) = (-t - \sin 5t, t^2 + \cos 6t) = (-x(t), y(t))$. This means that the graph will be symmetric about the y-axis and the only plot that remains with this property is F. What remains is to match the formulas in (d) and (e) with the graphs in B and C. This is easy: in (d) large positive values of t give points in the first quadrant. The graph in C has no points in the first quadrant.

5. The velocity is $\mathbf{x}'(t) = \left(-\sin(t-1), 3t^2, -\frac{1}{t^2}\right)$ so $\mathbf{x}'(1) = (0, 3, -1)$. At $t = 1$ the position is $\mathbf{x}(1) = (1, 0, -1)$. If we define a surface by the equation $f(x, y, z) = x^3 + y^3 + z^3 - xyz = 0$, then $\nabla f(x, y, z) = (3x^2 - yz, 3y^2 - xz, 3z^2 - xy)$ so $\nabla f(1, 0, -1) = (3, 1, 3)$. In general this vector is normal to the tangent plane at $(1, 0, -1)$ and by observation it is also perpendicular to $\mathbf{x}'(1)$ so the curve is tangent to the surface when $t = 1$.

7. For $w = 0$ we just get the line segment joining the points x_1 and x_2. As w increases the curve becomes more bent in the direction of the control point.

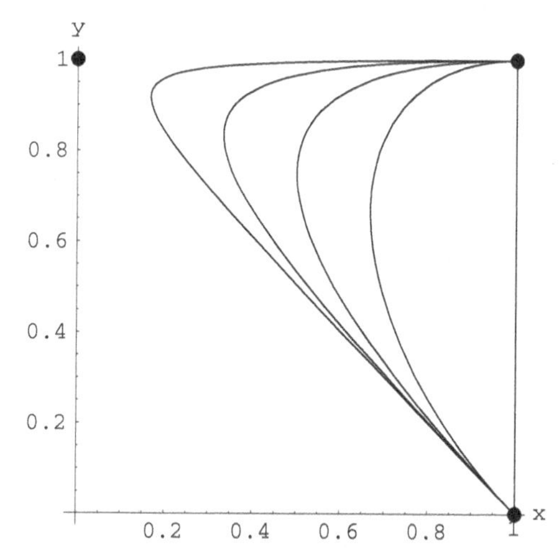

11. **(a)** Use part (b) of Exercise 9.

$$a = \|\mathbf{x}(1/2) - (x_2, y_2)\| = \left\| \frac{1}{1+w} \left(\frac{x_1 + x_3}{2} + wx_2, \frac{y_1 + y_3}{2} + wy_2 \right) - (x_2, y_2) \right\|$$

$$= \left\| \left(\frac{1}{2(1+w)} \right) (x_1 - 2x_2 + x_3, y_1 - 2y_2 + y_3) \right\|$$

$$= \left(\frac{1}{2(1+w)} \right) \sqrt{(x_1 - 2x_2 + x_3)^2 + (y_1 - 2y_2 + y_3)^2}$$

(b) This is a similar calculation.

$$\begin{aligned} b &= \left\| \mathbf{x}(1/2) - \left(\frac{x_1+x_3}{2}, \frac{y_1+y_3}{2}\right) \right\| \\ &= \left\| \frac{1}{1+w}\left(\frac{x_1+x_3}{2} + wx_2, \frac{y_1+y_3}{2} + wy_2\right) - \left(\frac{x_1+x_3}{2}, \frac{y_1+y_3}{2}\right) \right\| \\ &= \left\| \left(\frac{w}{2(1+w)}\right)(-x_1+2x_2-x_3, -y_1+2y_2-y_3) \right\| \\ &= \left(\frac{w}{2(1+w)}\right)\sqrt{(x_1-2x_2+x_3)^2+(y_1-2y_2+y_3)^2} \end{aligned}$$

(c) It's kind of amazing, but

$$\frac{b}{a} = \frac{\left(\frac{w}{2(1+w)}\right)\sqrt{(x_1-2x_2+x_3)^2+(y_1-2y_2+y_3)^2}}{\left(\frac{1}{2(1+w)}\right)\sqrt{(x_1-2x_2+x_3)^2+(y_1-2y_2+y_3)^2}} = w.$$

15. If $r = f(\theta)$, then we may write $\mathbf{x}(\theta) = (f(\theta)\cos\theta, f(\theta)\sin\theta)$. Hence $\mathbf{v} = \mathbf{x}'(\theta) = (f'(\theta)\cos\theta - f(\theta)\sin\theta, f'(\theta)\sin\theta + f(\theta)\cos\theta)$ and $\|\mathbf{v}\| = \sqrt{f'(\theta)^2 + f(\theta)^2} = \sqrt{r'^2 + r^2}$. Also

$$\mathbf{a} = \mathbf{x}''(\theta) = (f''(\theta)\cos\theta - 2f'(\theta)\sin\theta - f(\theta)\cos\theta, f''(\theta)\sin\theta + 2f'(\theta)\cos\theta - f(\theta)\sin\theta).$$

If we calculate $\mathbf{v} \times \mathbf{a}$, we find (after same algebra)

$$\mathbf{v} \times \mathbf{a} = (-f(\theta)f''(\theta) + 2f'(\theta)^2 + f(\theta)^2)\mathbf{k} = (r^2 - rr'' + 2r'^2)\mathbf{k}.$$

Hence, using formula (17), we have

$$\kappa = \frac{\|\mathbf{v} \times \mathbf{a}\|}{\|\mathbf{v}\|^3} = \frac{|r^2 - rr'' + 2r'^2|}{(r^2 + r'^2)^{3/2}}.$$

19. **(a)** This first conclusion is pretty much by definition. Analytically,

$$\|\mathbf{y}(t) - \mathbf{x}(t)\| = \|\mathbf{x}(t) - s(t)\mathbf{T}(t) - \mathbf{x}(t)\| = \|s(t)\mathbf{T}(t)\| = |s(t)|\|\mathbf{T}(t)\| = |s(t)| = s(t).$$

This last fact follows because $s(t) \geq 0$. Finally we note that $s(t)$ is the distance traveled from $\mathbf{x}(t_0)$ to $\mathbf{x}(t)$ along the underlying curve of $\mathbf{x}$.

(b) We calculated the distance from $\mathbf{x}$ to $\mathbf{y}$ in part (a). We should also observe that this is the distance along the tangent line to $\mathbf{x}$ at time t as it included the point $\mathbf{x}(t)$ and was in the direction $\mathbf{T}(t)$. The conclusion follows—it is as if you are unwinding a taut string from around $\mathbf{x}$: at each point $\mathbf{y}(t)$ is at a point in the direction of the tangent to $\mathbf{x}(t)$ of distance equal to the distance already traveled along $\mathbf{x}$. In other words, the distance is equal to the string already unraveled.

21. The curvature of a circle of radius a is $\kappa = 1/a$. Recall from high school geometry that a tangent to a circle at a given point is perpendicular to a radial line at that point. If the normal vector is oriented inward, then the evolute consists of points of distance equal to the radius of the circle in the direction of the center of the circle. In other words, the evolute of a circle is the center of that circle. Analytically, this is $\mathbf{e}(t) = \mathbf{x}(t) + a\mathbf{N}(t)$.

25. Assume that $\mathbf{x}$ is a unit speed curve. To get the direction of the tangent, consider $\mathbf{e}'(t) = \mathbf{x}'(t) + (-1/\kappa^2)\kappa'\mathbf{N}(t) + (1/\kappa)\mathbf{N}'(t)$. By the Frenet–Serret equations, $\mathbf{N}'(t) = -\kappa\mathbf{T}$ since $\mathbf{x}$ is a planar curve. So we see what remains is $\mathbf{e}'(t) = (-1/\kappa^2)\kappa'\mathbf{N}(t)$. This tells us that the unit tangent vector to the evolute is the parallel to the unit normal vector to the original path.

27. **(a)** $[x'(s)]^2 + [y'(s)]^2 = [\cos g(s)]^2 + [\sin g(s)]^2 = 1.$

(b) $\mathbf{v}(s) = (\cos g(s), \sin g(s))$ and $\mathbf{a}(s) = (-g'(s)\sin g(s), g'(s)\cos g(s))$ so

$$\kappa = \frac{\|\mathbf{v} \times \mathbf{a}\|}{\|\mathbf{v}\|^3} = \|(0, 0, g'(s))\| = |g'(s)|.$$

(c) We use the defining equations with $g'(s) = \kappa(s)$.

(d) There is more than one solution. For $s \geq 0$ we have $\kappa = s$ therefore, $g(s) = s^2/2$ so

$$x(s) = \int_0^s \cos(t^2/2)\,dt \quad \text{and} \quad y(s) = \int_0^s \sin(t^2/2)\,dt.$$

For $s < 0$, one solution corresponds to $g(s) = -s^2/2$ because for $s < 0, g'(s) = -s = |s|$. By formula (8) in Section 3.2, κ will always be non-negative, so we can also take $g(s) = s^2/2$ for $s < 0$. Because cosine is an even function and sine is an odd function, our two solutions are

$$x(s) = \int_0^s \cos(t^2/2)\,dt \quad \text{and} \quad y(s) = \pm \int_0^s \sin(t^2/2)\,dt.$$

(e) The graph of the clothoid is shown below.

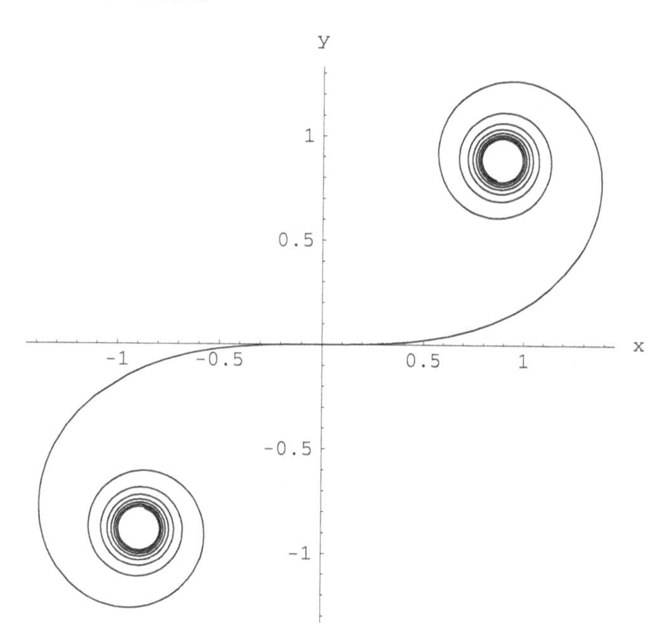

31. (a) The plan is to find the curvature of the strake by finding the curvature of the helix along the pipe. The radius r will be the reciprocal of this curvature (since the curvature of a circle of radius r is $1/r$). The path is $\mathbf{x}(t) = (a\cos\, t, a\sin\, t, ht/2\pi)$ so $\mathbf{x}'(t) = (-a\sin\, t, a\cos\, t, h/2\pi)$ and $\mathbf{x}''(t) = (-a\cos t, -a\sin t, 0)$.

$$\kappa = \frac{\|(-a\sin t, a\cos t, h)/2\pi \times (-a\cos t, -a\sin t, 0)\|}{\|(-a\sin t, a\cos t, h/2\pi)\|^3} = \frac{\|((ah/2\pi)\sin t, -(ah/2\pi)\cos t, a^2)\|}{(a^2 + [h/2\pi]^2)^{3/2}}$$

$$= \frac{a\sqrt{h^2/4\pi^2 + a^2}}{(a^2 + [h/2\pi]^2)^{3/2}} = \frac{a}{a^2 + h^2/4\pi^2} \quad \text{so} \quad r = \frac{(a^2 + h^2/4\pi^2)}{a}.$$

(b) If $a = 3$ and $h = 25$ then $r = (9 + 625/4\pi^2)/3 \approx 8.2771$.

35. $\mathbf{N}' = -\kappa\mathbf{T} + \tau\mathbf{B}$ by the Frenet–Serret formula. Now for $\mathbf{N}$ to be defined $\kappa \neq 0$, so if $\tau = 0$, then $\mathbf{N}' = -\kappa\mathbf{T} \neq \mathbf{0}$ (hence $\mathbf{N}$ is not constant). If $\tau \neq 0$, then for $\mathbf{N}'$ to be $\mathbf{0}$, $\mathbf{T}$ and $\mathbf{B}$ would have to be parallel, which they aren't.

39. $\mathbf{F} = u(x,y)\mathbf{i} - v(x,y)\mathbf{j}$ is an incompressible, irrotational vector field and so $\nabla \cdot \mathbf{F} = 0$ and $\nabla \times \mathbf{F} = \mathbf{0}$.

(a) The Cauchy–Riemann equations follow immediately from the assumptions:

$$0 = \nabla \cdot \mathbf{F} = \frac{\partial u}{\partial x} - \frac{\partial v}{\partial y}, \text{ so}$$

$$\frac{\partial u}{\partial x} = \frac{\partial v}{\partial y}, \text{ and}$$

$$0 = (\nabla \times \mathbf{F}) \cdot \mathbf{k} = -\frac{\partial v}{\partial x} - \frac{\partial u}{\partial y}, \text{ so}$$

$$\frac{\partial u}{\partial y} = -\frac{\partial v}{\partial x}$$

(b) Take the partial derivative with respect to x of both sides of the equation: $\dfrac{\partial u}{\partial x} = \dfrac{\partial v}{\partial y}$:

$$\frac{\partial^2 u}{\partial x^2} = \frac{\partial}{\partial x}\frac{\partial v}{\partial y} = \frac{\partial}{\partial y}\frac{\partial v}{\partial x} = -\frac{\partial}{\partial y}\frac{\partial u}{\partial y} = -\frac{\partial^2 u}{\partial y^2}.$$

An analogous calculation shows the result for v.

43. Notice that $\nabla \times \mathbf{F} = (0, 2e^{-x}\cos z, 0) \neq \mathbf{0}$. If $\mathbf{F}$ were a gradient field ∇f of class C^2, then, by Theorem 4.3, $\nabla \times \mathbf{F} = \nabla \times (\nabla f) = \mathbf{0}$.

Chapter 4

Maxima and Minima in Several Variables

4.1 Differentials and Taylor's Theorem

3. First calculate $f(x), f'(x), \ldots, f^{(k)}(x)$ and $f(a), f'(a), \ldots, f^{(k)}(a)$. Then we'll plug into the formula for Taylor's theorem in one variable (Theorem 1.1 in the text) with $a = 1$ and $k = 4$:

$$\begin{aligned} f(x) &= \frac{1}{x^2} & f(1) &= 1 \\ f'(x) &= -\frac{2}{x^3} & f'(1) &= -2 \\ f''(x) &= \frac{6}{x^4} & f''(1) &= 6 \\ f'''(x) &= -\frac{24}{x^5} & f'''(1) &= -24 \\ f''''(x) &= \frac{120}{x^6} & f''''(1) &= 120, \end{aligned}$$

so

$$\begin{aligned} p_4(x) &= 1 - 2(x-1) + \frac{6}{2}(x-1)^2 - \frac{24}{6}(x-1)^3 + \frac{120}{24}(x-1)^4 \\ &= 1 - 2(x-1) + 3(x-1)^2 - 4(x-1)^3 + 5(x-1)^4. \end{aligned}$$

7. Here $a = \pi/2$ and $k = 5$, so

$$\begin{aligned} f(x) &= \sin x & f(\pi/2) &= 1 \\ f'(x) &= \cos x & f'(\pi/2) &= 0 \\ f''(x) &= -\sin x & f'(\pi/2) &= -1 \\ f'''(x) &= -\cos x & f'(\pi/2) &= 0 \\ f''''(x) &= \sin x & f'(\pi/2) &= 1 \\ f'''''(x) &= \cos x & f'(\pi/2) &= 0. \end{aligned}$$

Therefore,

$$p_5(x) = 1 - \frac{(x-\pi/2)^2}{2} + \frac{(x-\pi/2)^4}{24}.$$

11. Here $\mathbf{a} = (0, \pi)$ and

$$\begin{aligned} f(x,y) &= e^{2x}\cos 3y & f(0,\pi) &= -1 \\ f_x(x,y) &= 2e^{2x}\cos 3y & f_x(0,\pi) &= -2 \\ f_y(x,y) &= -3e^{2x}\sin 3y & f_y(0,\pi) &= 0 \\ f_{xx}(x,y) &= 4e^{2x}\cos 3y & f_{xx}(0,\pi) &= -4 \\ f_{yy}(x,y) &= -9e^{2x}\cos 3y & f_{yy}(0,\pi) &= 9 \\ f_{xy}(x,y) &= -6e^{2x}\sin 3y & f_{xy}(0,\pi) &= 0, \end{aligned}$$

so

$$p_1(\mathbf{x}) = -1 - 2x \quad \text{and}$$
$$p_2(\mathbf{x}) = -1 - 2x + \frac{1}{2}(-4x^2 + 9(y-\pi)^2)$$
$$= -1 - 2x - 2x^2 + \frac{9}{2}(y-\pi)^2.$$

15. Here $\mathbf{a} = (0,0,0)$ and there is quite a bit of symmetry so we'll only calculate:

$$\begin{aligned} f(x,y,z) &= \sin xyz & f(0,0,0) &= 0 \\ f_x(x,y,z) &= yz\cos xyz & f_x(0,0,0) &= 0 = f_y(0,0,0) = f_z(0,0,0) \\ f_{xx}(x,y,z) &= -y^2z^2\sin xyz & f_{xx}(0,0,0) &= 0 = f_{yy}(0,0,0) = f_{zz}(0,0,0) \\ f_{xy}(x,y) &= z\cos xyz - xyz^2\sin xyz & f_{xy}(0,0,0) &= 0 = f_{xz}(0,0,0) = f_{yz}(0,0,0) \end{aligned}$$

$$\text{so} \quad p_1(\mathbf{x}) = 0 \quad \text{and} \quad p_2(\mathbf{x}) = 0.$$

19. $f(x,y,z) = x^3 + x^2y - yz^2 + 2z^3$

$$\begin{array}{lll} f_x(x,y,z) = 3x^2 + 2xy & f_y(x,y,z) = x^2 - z^2 & f_z(x,y,z) = -2yz + 6z^2 \\ f_{xx}(x,y,z) = 6x + 2y & f_{yx}(x,y,z) = 2x & f_{zx}(x,y,z) = 0 \\ f_{xy}(x,y,z) = 2x & f_{yy}(x,y,z) = 0 & f_{zy}(x,y,z) = -2z \\ f_{xz}(x,y,z) = 0 & f_{yz}(x,y,z) = -2z & f_{zz}(x,y,z) = -2y + 12z \end{array}$$

so

$$Hf(1,0,1) = \begin{bmatrix} 6 & 2 & 0 \\ 2 & 0 & -2 \\ 0 & -2 & 12 \end{bmatrix}.$$

21. Use the work from Exercises 8 and 16:

$$p_2(\mathbf{x}) = f(0,0) + Df(0,0)\mathbf{x} + \frac{1}{2}\mathbf{x}^T \begin{bmatrix} -2 & 0 \\ 0 & -2 \end{bmatrix} \mathbf{x}$$
$$= 1 + \frac{1}{2}\begin{bmatrix} x & y \end{bmatrix} \begin{bmatrix} -2 & 0 \\ 0 & -2 \end{bmatrix} \begin{bmatrix} x \\ y \end{bmatrix}.$$

25. The function is $f(x_1, x_2, \ldots, x_n) = e^{x_1 + 2x_2 + \cdots + nx_n}$.

(a) $Df(x_1, x_2, \ldots, x_n) = e^{x_1 + 2x_2 + \cdots + nx_n}\begin{bmatrix} 1 & 2 & \cdots & n \end{bmatrix}$, and therefore $Df(0,0,\ldots,0) = \begin{bmatrix} 1 & 2 & \cdots & n \end{bmatrix}$. Taking second derivatives and evaluating at the origin results in:

$$Hf(0,0,\ldots,0) = \begin{bmatrix} 1 & 2 & 3 & \cdots & n \\ 2 & 4 & 6 & \cdots & 2n \\ 3 & 6 & 9 & \cdots & 3n \\ \vdots & \vdots & \vdots & \ddots & \vdots \\ n & 2n & 3n & \cdots & n^2 \end{bmatrix}.$$

(c) Since (c) follows immediately from (a) we will skip (b) for a moment.

$$p_2(\mathbf{x}) = f(0,0,\ldots,0) + Df(0,0,\ldots,0)\mathbf{x} + \frac{1}{2}\mathbf{x}^T Hf(0,0,\ldots,0)\mathbf{x}$$

$$= 1 + \begin{bmatrix} 1 & 2 & \cdots & n \end{bmatrix} \begin{bmatrix} x_1 \\ x_2 \\ \vdots \\ x_n \end{bmatrix} + \frac{1}{2}\begin{bmatrix} x_1 & x_2 & \cdots & x_n \end{bmatrix} \begin{bmatrix} 1 & 2 & 3 & \cdots & n \\ 2 & 4 & 6 & \cdots & 2n \\ 3 & 6 & 9 & \cdots & 3n \\ \vdots & \vdots & \vdots & \ddots & \vdots \\ n & 2n & 3n & \cdots & n^2 \end{bmatrix} \begin{bmatrix} x_1 \\ x_2 \\ \vdots \\ x_n \end{bmatrix}.$$

(b) Now we can read the answer to (b) right off of our answer to (c).

$$p_1(\mathbf{x}) = 1 + x_1 + 2x_2 + \cdots + nx_n \quad \text{and}$$

$$p_2(\mathbf{x}) = 1 + x_1 + 2x_2 + \cdots + nx_n + \frac{1}{2}\sum_{i,j=1}^{n} ijx_ix_j.$$

27. $Df(x,y,z) = \begin{bmatrix} 4x^3 + 3x^2y - z^2 + 2xy + 3y & x^3 + 6y^2 + x^2 + 3x & -2xz - 1 \end{bmatrix}$ and

$$Hf(x,y,z) = \begin{bmatrix} 12x^2 + 6xy + 2y & 3x^2 + 2x + 3 & -2z \\ 3x^2 + 2x + 3 & 12y & 0 \\ -2z & 0 & -2x \end{bmatrix}.$$

The only non-zero third derivatives are

$$f_{xxx}(x,y,z) = 24x + 6y \qquad f_{xxy}(x,y,z) = 6x + 2$$
$$f_{xzz}(x,y,z) = -2 \qquad f_{yyy}(x,y,z) = 12$$

and their permutations.

(a) Here $\mathbf{a} = (0,0,0)$ so $f(0,0,0) = 2$, $Df(0,0,0) = \begin{bmatrix} 0 & 0 & -1 \end{bmatrix}$, and $Hf(0,0,0) = \begin{bmatrix} 0 & 3 & 0 \\ 3 & 0 & 0 \\ 0 & 0 & 0 \end{bmatrix}$.

$$p_3(\mathbf{x}) = 2 - z + 3xy + \frac{1}{6}(6x^2y - 6xz^2 + 12y^3)$$
$$= 2 - z + 3xy + x^2y - xz^2 + 2y^3.$$

(b) Here $f(1,-1,0) = -4$, $Df(1,-1,0) = \begin{bmatrix} -4 & 11 & -1 \end{bmatrix}$, and $Hf(1,-1,0) = \begin{bmatrix} 4 & 8 & 0 \\ 8 & -12 & 0 \\ 0 & 0 & -2 \end{bmatrix}$.

$$\begin{aligned} p_3(\mathbf{x}) &= -4 - 4(x-1) + 11(y+1) - z \\ &\quad + \frac{1}{2}[4(x-1)^2 + 16(x-1)(y+1) - 12(y+1)^2 - 2z^2] \\ &\quad + \frac{1}{6}[18(x-1)^3 + 3(8)(x-1)^2(y+1) - 3(2)(x-1)z^2 + 12(y+1)^3] \\ &= -4 - 4(x-1) + 11(y+1) - z + 2(x-1)^2 + 8(x-1)(y+1) - 6(y+1)^2 - z^2 \\ &\quad + 3(x-1)^3 + 4(x-1)^2(y+1) - (x-1)z^2 + 2(y+1)^3. \end{aligned}$$

31. $f(x,y,z) = e^x \cos y + e^y \sin z$ so $df(x,y,z,\mathbf{h}) = e^x \cos y\, dx + (-e^x \sin y + e^y \sin z)\, dy + e^y \cos z\, dz$.

35. Although you will probably solve this more formally, you should see that, intuitively, changes in the upper left entry are multiplied by the largest number so that is the entry for which the value of the determinant is most sensitive.

39. We are told that $dr = dh$ and know that $V = (1/3)\pi r^2 h$. So $dV = (2/3)\pi rh\, dr + (1/3)\pi r^2 dh = (28\pi/3)\, dr$. Now we want $|dV|$ to be at most .2 so $|dV| = (28\pi/3)|dr| \le .2$ or $|dr| \le .3/(14\pi) \approx .0068209$.

43. (a) The preliminary calculations for $f(x,y) = e^{2x}\cos y$ are

$$\begin{aligned} f(x,y) &= e^{2x}\cos y & f(0,\pi/2) &= 0 \\ f_x(x,y) &= 2e^{2x}\cos y & f_x(0,\pi/2) &= 0 \\ f_y(x,y) &= -e^{2x}\sin y & f_y(0,\pi/2) &= -1 \\ f_{xx}(x,y) &= 4e^{2x}\cos y & f_{xx}(0,\pi/2) &= 0 \\ f_{xy}(x,y) &= -2e^{2x}\sin y & f_{xy}(0,\pi/2) &= -2 = f_{yx}(0,\pi/2) \\ f_{yy}(x,y) &= -e^{2x}\cos y & f_{yy}(0,\pi/2) &= 0. \end{aligned}$$

Thus

$$p_2(x,y,z) = -\left(y - \frac{\pi}{2}\right) + \frac{1}{2}\left(-4x\left(y - \frac{\pi}{2}\right)\right) = \frac{\pi}{2} - y - 2x\left(y - \frac{\pi}{2}\right).$$

(b) The eight third-order partial derivatives are:

$$\begin{aligned} f_{xxx}(x,y) &= 8e^{2x}\cos y \\ f_{xxy}(x,y) &= -4e^{2x}\sin y = f_{xyx}(x,y) = f_{yxx}(x,y) \\ f_{xyy}(x,y) &= -2e^{2x}\cos y = f_{yxy}(x,y) = f_{yyx}(x,y) \\ f_{yyy}(x,y) &= e^{2x}\sin y, \end{aligned}$$

Lagrange's form of the remainder tells us that

$$\left|R_2\left(x,y,0,\frac{\pi}{2}\right)\right| = \frac{1}{3!}\left|\sum_{i,j,k=1}^{2} f_{x_ix_jx_k}(\mathbf{z})h_ih_jh_k\right|,$$

where $\mathbf{z}$ is a point on the line segment joining $(0,\pi/2)$ and (x,y). Note that the exponential function e^{2x} increases with x and the sine and cosine have maximum values of 1. Thus

$$|f_{xxx}(x,y)| \le 8e^{0.4},$$

and similar results apply to the other third-order partials. Hence

$$\begin{aligned} \left|R_2\left(x,y,0,\frac{\pi}{2}\right)\right| &\le \frac{1}{6}\left(8e^{0.4}|h_1|^3 + 3\cdot 4e^{0.4}|h_1|^2|h_2| + 3\cdot 2e^{0.4}|h_1|\,|h_2|^2 + e^{0.4}|h_2|^3\right) \\ &= \frac{e^{0.4}}{6}\left(8|h_1|^3 + 12|h_1|^2|h_2| + 6|h_1|\,|h_2|^2 + |h_2|^3\right). \end{aligned}$$

If $|h_1| \le 0.2$ and $|h_2| \le 0.1$, then

$$\left|R_2\left(x,y,0,\frac{\pi}{2}\right)\right| \le \frac{e^{0.4}}{6}\left(8(0.008) + 12(0.004) + 6(0.002) + 0.001\right) \approx 0.03108.$$

4.2 Extrema of Functions

1. $f(x,y) = 4x + 6y - 12 - x^2 - y^2$ so $f_x(x,y) = 4 - 2x$, $f_y(x,y) = 6 - 2y$, $f_{xx}(x,y) = -2$, $f_{xy}(x,y) = 0$, and $f_{yy}(x,y) = -2$.

(a) To find the critical point we will set each of the first partial derivatives equal to 0 and solve: $f_x(x,y) = 0$ when $4-2x = 0$ or when $x = 2$ and $f_y(x,y) = 0$ when $6 - 2y = 0$ or when $y = 3$. So f has a unique critical point at (2, 3).

(b) The increment

$$\begin{aligned} \Delta f &= f(2+\Delta x, 3+\Delta y) - f(2,3) \\ &= 4(2+\Delta x) + 6(3+\Delta y) - 12 - (2+\Delta x)^2 - (3+\Delta y)^2 \\ &\quad - (4(2) + 6(3) - 12 - 2^2 - 3^2) = -(\Delta x)^2 - (\Delta y)^2. \end{aligned}$$

This tells us that little changes in x and/or y result in a decrease in the value of f. This means that f must have a local maximum at (2, 3).

(c) The Hessian is $Hf(2,3) = \begin{bmatrix} -2 & 0 \\ 0 & -2 \end{bmatrix}$ so $d_1 = -2$ and $d_2 = 4$ so by the second derivative test, f has a local maximum at (2, 3).

5. $f(x,y) = x^2 + y^3 - 6xy + 3x + 6y$, so $f_x(x,y) = 2x - 6y + 3$ and $f_y(x,y) = 3y^2 - 6x + 6$. At a critical point for f, $2x = 6y - 3$ and $0 = 3y^2 - 6x + 6$ so $0 = y^2 - 2x + 2$. Substituting, $0 = y^2 - 6y + 5 = (y-1)(y-5)$. We have critical points at (3/2, 1) and (27/2, 5).
The second derivatives are $f_{xx}(x,y) = 2$, $f_{yy}(x,y) = 6y$, and $f_{xy}(x,y) = -6$. $d_1 = 2$ and $d_2 = 12y - 36$. In other words, d_1 is always positive and d_2 is positive when $y = 5$ and negative when $y = 1$ so by the second derivative test f has a saddle point at (3/2, 1) and f has a local minimum at (27/2, 5).

11. $f(x, y) = x^2 - y^3 - x^2y + y$, so $f_x(x, y) = 2x - 2xy = 2x(1 - y)$ and $f_y(x, y) = -3y^2 - x^2 + 1$. At a critical point for f, either $x = 0$ or $y = 1$. When $x = 0$, y must be $\pm 1/\sqrt{3}$. No solution corresponds to $y = 1$, So the critical points for f are $(0, \pm 1/\sqrt{3})$.
The second derivatives are $f_{xx}(x, y) = 2 - 2y$, $f_{yy}(x, y) = -6y$, and $f_{xy}(x, y) = -2x$. $d_1 = 2 - 2y$ and $d_2 = -12y + 12y^2 - 4x^2$. At $(0, -1/\sqrt{3})$, d_1 is positive and d_2 is positive so f has a local minimum at $(0, -1/\sqrt{3})$. At $(0, 1/\sqrt{3})$, d_1 is positive and d_2 is negative so f has a saddle point at $(0, 1/\sqrt{3})$.

15. $f(x, y, z) = x^2 - xy + z^2 - 2xz + 6z$, so $f_x(x, y, z) = 2x - y - 2z$, $f_y(x, y, z) = -x$ and $f_z(x, y, z) = 2z - 2x + 6$. From the second equation, $x = 0$. From the third, then, $z = -3$ and from the first it follows that $y = 6$.
The second derivatives are $f_{xx}(x, y, z) = 2$, $f_{yy}(x, y, z) = 0$, $f_{zz}(x, y, z) = 2$, $f_{xy}(x, y, z) = -1$, $f_{xz}(x, y, z) = -2$ and $f_{yz}(x, y, z) = 0$. $d_1 = 2$, $d_2 = -1$ and $d_3 = -2$ so f has a saddle point at $(0, 6, -3)$.

19. $f(x, y, z) = xy + xz + 2yz + \dfrac{1}{x}$ so $f_x(x, y, z) = y + z - \dfrac{1}{x^2}$, $f_y(x, y, z) = x + 2z$, and $f_z(x, y, z) = x + 2y$. We see immediately that at a critical point of f, $y = z$ so both $2z = -x$ and $2z = \dfrac{1}{x^2}$ so $-x = \dfrac{1}{x^2}$ so $x = -1$. Therefore, f has a critical point at $(-1, 1/2, 1/2)$.
The Hessian is $Hf = \begin{bmatrix} 2/x^3 & 1 & 1 \\ 1 & 0 & 2 \\ 1 & 2 & 0 \end{bmatrix}$ so $d_1(-1, 1/2, 1/2) = -2$, $d_2(-1, 1/2, 1/2) = -1$, and $d_3(-1, 1/2, 1/2) = 12$.
This is the case of the second derivative test where the conditions are valid but neither of the first two cases holds so f has a saddle point at $(-1, 1/2, 1/2)$.

23. If you think of this problem geometrically it should be reasonably straightforward. The slices through the origin where only one variable is allowed to change are parabolas. They open up if the coefficient of the term containing that variable is positive and down if it is negative. This tells you that if all of the coefficients are positive then we have a local minimum, if all of the coefficients are negative then we have a local maximum, and if some are positive and some are negative then we have a saddle point.

(a) $f(x, y) = ax^2 + by^2$ so $f_x(x, y) = 2ax$ and $f_y(x, y) = 2by$. Since neither a nor b is 0, the critical point must be the origin. The Hessian is $Hf = \begin{bmatrix} 2a & 0 \\ 0 & 2b \end{bmatrix}$. The first condition is that $d_2 > 0$ so $4ab > 0$ so a and b are the same sign. Also, $d_1 = 2a$ so when a and b are negative the origin is a local maximum and when a and b are positive the origin is a local minimum.

(b) $f(x, y) = ax^2 + by^2 + cz^2$ so $f_x(x, y, z) = 2ax$, $f_y(x, y, z) = 2by$ and $f_z(x, y, z) = 2cz$. Since none of a, b and c is 0, the critical point must be the origin. The Hessian is $Hf = \begin{bmatrix} 2a & 0 & 0 \\ 0 & 2b & 0 \\ 0 & 0 & 2c \end{bmatrix}$. Again, in either case $d_2 > 0$ so $4ab > 0$ so a and b are the same sign. Also, $d_1 = 2a$ and $d_3 = 8abc$. In either case d_1 and d_3 must be the same sign. When a, b and c are negative the origin is a local maximum and when a, b and c are positive the origin is a local minimum.

(c) Really the analysis is no harder, it is just harder to write down. The function is now $f(x_1, x_2, \ldots, x_n) = a_1x_1^2 + a_2x_2^2 + \cdots + a_nx_n^2$. The first derivatives are $f_{x_i}(x_1, x_2, \ldots, x_n) = 2a_ix_i$. Because none of the a_i is zero and all of the first derivatives are 0, we conclude that the only critical point is at the origin. The Hessian is an $n \times n$ matrix with zeros everywhere off of the main diagonal and the entry in position (i, i) is $2a_i$. We easily calculate $d_i = 2^ia_1a_2 \ldots a_i$. As above, d_2 must be positive so both a_1 and a_2 are of the same sign. We could continue to argue that $d_4 = 4a_3a_4d_2$ so a_3 and a_4 must be of the same sign. In fact, we can continue that reasoning to say for k odd, a_k and a_{k+1} must be of the same sign. For f to have a local maximum $d_1 < 0$ so a_1 and a_2 are both negative. Also, $d_k = 2a_kd_{k-1}$ and for k odd $d_k < 0$ so we can move up through the entries and argue that all of the a_i's must be negative. Similarly, for f to have a local minimum all of the a_i's must be positive.

27. For *Mathematica*, we'll use the following sequence of commands:

- $f[x\_, y\_, z\_, w\_] = y\,w - x\,y\,z - x^2 - 2z^2 + w^2$

- Solve $[\{D[f[x, y, z, w], x] == 0, D[f[x, y, z, w], y] == 0,$

  $D[f[x, y, z, w], z] == 0, D[f[x, y, z, w], w] == 0\}]$

- The Hessian

$$\begin{aligned} H =&\{\{\partial_{x,x}f[x,y,z,w], \partial_{x,y}f[x,y,z,w], \partial_{x,z}f[x,y,z,w], \partial_{x,w}f[x,y,z,w]\},\\ &\{\partial_{y,x}f[x,y,z,w], \partial_{y,y}f[x,y,z,w], \partial_{y,z}f[x,y,z,w], \partial_{y,w}f[x,y,z,w]\},\\ &\{\partial_{z,x}f[x,y,z,w], \partial_{z,y}f[x,y,z,w], \partial_{z,z}f[x,y,z,w], \partial_{z,w}f[x,y,z,w]\}\\ &\{\partial_{w,x}f[x,y,z,w], \partial_{w,y}f[x,y,z,w], \partial_{w,z}f[x,y,z,w], \partial_{w,w}f[x,y,z,w]\}\end{aligned}$$

- MatrixForm[$H/.\{x \to 0, y \to 0, z \to 0, w \to 0\}$] (since $(0,0,0,0)$ is a critical point found in the found second step)

The critical points are $(0,0,0,0), (-\sqrt{2}, 2\sqrt{2}, 1, -\sqrt{2}), (\sqrt{2}, 2\sqrt{2}, -1, -\sqrt{2}), (-\sqrt{2}, -2\sqrt{2}, -1, \sqrt{2})$, and $(\sqrt{2}, -2\sqrt{2}, 1, \sqrt{2})$.

At $(0,0,0,0)$ the Hessian is $\begin{bmatrix} -2 & 0 & 0 & 0 \\ 0 & 0 & 0 & 1 \\ 0 & 0 & -4 & 0 \\ 0 & 1 & 0 & 2 \end{bmatrix}$. So $d_1 = -2 < 0$, $d_2 = 0$, $d_3 = 0$, and $d_4 = -8 < 0$, so $(0,0,0,0)$ is a saddle point.

At $(-\sqrt{2}, 2\sqrt{2}, 1, -\sqrt{2})$ the Hessian is $\begin{bmatrix} -2 & -1 & -2\sqrt{2} & 0 \\ -1 & 0 & \sqrt{2} & 1 \\ -2\sqrt{2} & \sqrt{2} & -4 & 0 \\ 0 & 1 & 0 & 2 \end{bmatrix}$. So $d_1 = -2 < 0$, $d_2 = -1 < 0$, $d_3 = 16 > 0$, and $d_4 = 32 > 0$, so $(-\sqrt{2}, 2\sqrt{2}, 1, -\sqrt{2})$ is a saddle point.

At $(\sqrt{2}, 2\sqrt{2}, -1, -\sqrt{2})$ the Hessian is $\begin{bmatrix} -2 & 1 & -2\sqrt{2} & 0 \\ 1 & 0 & -\sqrt{2} & 1 \\ -2\sqrt{2} & -\sqrt{2} & -4 & 0 \\ 0 & 1 & 0 & 2 \end{bmatrix}$. So $d_1 = -2 < 0$, $d_2 = -1 < 0$, $d_3 = 16 > 0$, and $d_4 = 32 > 0$, so $(\sqrt{2}, 2\sqrt{2}, -1, -\sqrt{2})$ is a saddle point.

At $(-\sqrt{2}, -2\sqrt{2}, -1, \sqrt{2})$ the Hessian is $\begin{bmatrix} -2 & 1 & 2\sqrt{2} & 0 \\ 1 & 0 & \sqrt{2} & 1 \\ 2\sqrt{2} & \sqrt{2} & -4 & 0 \\ 0 & 1 & 0 & 2 \end{bmatrix}$. So $d_1 = -2 < 0$, $d_2 = -1 < 0$, $d_3 = 16 > 0$, and $d_4 = 32 > 0$, so $(-\sqrt{2}, -2\sqrt{2}, -1, \sqrt{2})$ is a saddle point.

At $(\sqrt{2}, -2\sqrt{2}, 1, \sqrt{2})$ the Hessian is $\begin{bmatrix} -2 & -1 & 2\sqrt{2} & 0 \\ -1 & 0 & -\sqrt{2} & 1 \\ 2\sqrt{2} & -\sqrt{2} & -4 & 0 \\ 0 & 1 & 0 & 2 \end{bmatrix}$. So $d_1 = -2 < 0$, $d_2 = -1 < 0$, $d_3 = 16 > 0$, and $d_4 = 32 > 0$, so $(\sqrt{2}, -2\sqrt{2}, 1, \sqrt{2})$ is a saddle point.

33. $f(x,y,z) = x^2 + xz - y^2 + 2z^2 + xy + 5x$ so $f_x(x,y,z) = 2x + y + z + 5$, $f_y(x,y,z) = x - 2y$, and $f_z(x,y,z) = x + 4z$. At a critical point for f, $x = 2y = -4z$ so $-5 = 2x + y + z = -8z - 2z + z = -9z$. Our only critical point is $(-20/9, -10/9, 5/9)$ which is not within our region. We need to check the value of f along the boundary of the region $-5 \le x \le 0, 0 \le y \le 3, 0 \le z \le 2$. This consists of six two-dimensional faces, twelve one-dimensional edges and eight vertices.

- $f(x,0,0) = x^2 + 5x$ has a minimum of -6.25 at $x = -5/2$ and a maximum of 0 at $x = -5$ or 0,
- $f(x,0,2) = x^2 + 7x + 8$ has a minimum of -4.25 at $x = -7/2$ and a maximum of 8 at $x = 0$,
- $f(x,3,0) = x^2 + 8x - 9$ has a minimum of 25 at $x = -4$ and a maximum of -9 at $x = 0$,
- $f(x,3,2) = x^2 + 10x - 1$ has a minimum of -26 at $x = -5$ and a maximum of -1 at $x = 0$,
- $f(-5,y,0) = -y^2 - 5y$ has a minimum of -24 at $y = 3$ and a maximum of 0 at $y = 0$,
- $f(0,y,0) = -y^2$ has a minimum of -9 at $y = 3$ and a maximum of 0 at $y = 0$,
- $f(-5,y,2) = -y^2 - 5y - 2$ has a minimum of -26 at $y = 3$ and a maximum of -2 at $y = 0$,
- $f(0,y,2) = 8 - y^2$ has a minimum of -1 at $y = 3$ and a maximum of 8 at $y = 0$,
- $f(-5,0,z) = 2z^2 - 5z$ has a minimum of $-25/8$ at $z = 5/4$ and a maximum of 0 at $z = 0$,
- $f(0,0,z) = 2z^2$ has a minimum of 0 at $z = 0$ and a maximum of 8 at $z = 2$,

- $f(-5, 3, z) = 2z^2 - 5z - 24$ has a minimum of $-217/8$ at $z = 5/4$ and a maximum of -24 at $z = 0$,
- $f(0, 3, z) = 2z^2 - 9$ has a minimum of -9 at $z = 0$ and a maximum of -1 at $z = 2$.

You also must check for extrema on each face and at each vertex. When you do you find: The absolute maximum is 8 at (0, 0, 2) and the absolute minimum is $-191/7$ at $(-32/7, 3, 8/7)$.

37. $f(x, y) = 2x^2 - 2xy + y^2 - y + 3$, so $f_x(x, y) = 4x - 2y$ and $f_y(x, y) = -2x + 2y - 1$. At a critical point for f we have $y = 2x$, so $-2x + 4x - 1 = 0$. Thus the only critical point is $(\frac{1}{2}, 1)$.

Now we need to consider the boundary of the region. It consists of three parts: (1) the horizontal line $y = 0$, where $0 \le x \le 2$; (2) the vertical line $x = 0$, where $0 \le y \le 2$; (3) the line $x + y = 2$ (or $y = 2 - x$), where $0 \le x \le 2$. Thus we compare

- $f(\frac{1}{2}, 1) = \frac{5}{2}$,
- $f(x, 0) = 2x^2 + 3$ has a minimum of 3 at $x = 0$ and a maximum of 11 at $x = 2$,
- $f(0, y) = y^2 - y + 3$ has a minimum of $\frac{11}{4}$ at $y = \frac{1}{2}$ and a maximum of 5 at $y = 2$,
- $f(x, 2 - x) = 5x^2 - 7x + 5$ has a minimum of $\frac{51}{20}$ at $x = \frac{7}{10}$ and a maximum of 11 at $x = 2$

Thus the absolute minimum is $\frac{5}{2}$ occurring at $(\frac{1}{2}, 1)$ and the absolute maximum is 11 occurring at $(2, 0)$.

39. The boundary of the closed ball is given by $x^2 + y^2 - 2y + z^2 + 4z = 0$. Completing the square, we find $x^2 + y^2 - 2y + 1 + z^2 + 4z + 4 = 5$ or $x^2 + (y - 1)^2 + (z + 2)^2 = 5$. (Note also that $x^2 + y^2 - 2y + z^2 + 4z = x^2 + (y - 1)^2 + (z + 2)^2 - 5$.)

The function $f(x, y, z) = e^{1-x^2-y^2+2y-z^2-4z}$ has

$$\begin{aligned} f_x(x, y, z) &= -2xe^{1-x^2-y^2+2y-z^2-4z} = 0 && \text{when } x = 0 \\ f_y(x, y, z) &= (-2y + 2)e^{1-x^2-y^2+2y-z^2-4z} = 0 && \text{when } y = 1 \\ f_z(x, y, z) &= (-2z - 4)e^{1-x^2-y^2+2y-z^2-4z} = 0 && \text{when } z = -2 \end{aligned}$$

So $(0, 1, -2)$ is an interior critical point (the only one). Note that on the boundary $x^2 + y^2 - 2y + z^2 + 4z = 0$, we have

$$f(x, y, z) = e^{1-0} = e$$

$$f(0, 1, -2) = e^{1-(-5)} = e^6 \leftarrow \text{so absolute max is at } (0, 1, -2).$$

The absolute minimum of e occurs at *all* points of the boundary. If we set $w = x^2+y^2-2y+z^2+4z$, then $f(x, y, z) = e^{1-w}$, so that it's clear that the minimum must occur when $w = 0$ (since $w \le 0$ defines the domain we are to consider). Likewise, the maximum must occur at the center of the ball.

43. $f(x, y, z) = x^2y^3z^4$: in every deleted neighborhood of the origin where $y > 0$, $f(x, y, z) > 0$; when $y < 0$, $f(x, y, z) < 0$ so f has neither a minimum nor a maximum at the origin.

45. $f(x, y, z) = 2 - x^4y^4 - z^4$: in every deleted neighborhood of the origin $x^4y^4 + z^4 > 0$ so $f(x, y, z) < 2$ so $f(0, 0, 0) > f(x, y, z)$ for every point (x, y, z) near but not equal to (0, 0, 0) so f has a local maximum at the origin.

49. There can't be a global maximum because, for example, for fixed y, as $x \to 0+$ the function grows without bound. $f_x(x, y) = y - 1/x$ and $f_y(x, y) = x + 2 - 2/y$ so f has a critical point at (2, 1/2). From the Hessian $\begin{bmatrix} 1/4 & 1 \\ 1 & 8 \end{bmatrix}$ we see that there is a local minimum at (2, 1/2) of $2 + \ln 2$. Note that $f_x(2, y) = y - 1/2$.

We would like to now conclude that f has a unique critical point at (2, 1/2) which is a local minimum and hence it is a global minimum—such a conclusion seems reasonable, but, as Exercise 52 will demonstrate, is not correct. Consider $f_x(x, 1/2) = 1/2 - 1/x$. For $x > 2$ this is positive and so f is increasing along this line. Now look at $f_y(x, y) = x + 2 - 2/y$ for $x \ge 2$. When $y > 1/2$ this is positive and when $0 < y < 1/2$ this is negative. So as we move vertically away from the line $y = 1/2$ for $x \ge 2$ we see that f is increasing. A similar analysis for the remaining regions shows that f has a global minimum at (2, 1/2).

4.3 Lagrange Multipliers

3. The function is $f(x, y) = 5x + 2y$ subject to the constraint $g(x, y) = 5x^2 + 2y^2 = 14$. We solve the system

$$\begin{cases} 5 = 10\lambda x \\ 2 = 4\lambda y \\ 5x^2 + 2y^2 = 14. \end{cases}$$

By either of the first two equations we see that $\lambda \ne 0$. Together, the first two equations imply that $x = y$ so $7x^2 = 14$ so the critical points are $\pm(\sqrt{2}, \sqrt{2})$.

7. The function is $f(x,y,z) = 3 - x^2 - 2y^2 - z^2$ subject to the constraint $g(x,y,z) = 2x + y + z = 2$. We solve the system

$$\begin{cases} -2x = 2\lambda \\ -4y = \lambda \\ -2z = \lambda \\ 2x + y + z = 2. \end{cases}$$

Immediately we have $\lambda = -x = -4y = -2z \iff x = 4y = 2z$. Thus $x = 2z$ and $y = z/2$ so that the last equation of the system becomes $4z + z/2 + z = 2 \iff z = 4/11$. Therefore, there is a unique critical point of $\left(\frac{8}{11}, \frac{2}{11}, \frac{4}{11}\right)$.

13. **(a)** The function is $f(x,y) = x^2 + y$ subject to the constraint $g(x,y) = x^2 + 2y^2 = 1$. We solve the system

$$\begin{cases} 2x = 2x\lambda \\ 1 = 4y\lambda \\ x^2 + 2y^2 = 1. \end{cases}$$

From the first equation, we see that either $x = 0$ or $\lambda = 1$. If $\lambda = 1$, then $y = 1/4$, so $x = \pm\sqrt{7/8}$. If $x = 0$, then $y = \pm\sqrt{1/2}$. In short, the critical points are $(\pm\sqrt{7/8}, 1/4)$ and $(0, \pm\sqrt{1/2})$.

(b) $L(\lambda; x, y) = x^2 + y - \lambda(x^2 + 2y^2 - 1)$ so

$$H(\lambda; x, y) = \begin{bmatrix} 0 & -2x & -4y \\ -2x & 2 - 2\lambda & 0 \\ -4y & 0 & -4\lambda \end{bmatrix}.$$

So $-d_3 = -16y[x^2 + 1/2 - 2y]$. Substitute the critical points to find that there are local maxima at $(\pm\sqrt{7/8}, 1/4)$ and local minima at $(0, \pm\sqrt{1/2})$.

15. Input the following three lines into *Mathematica* (or the equivalent into your favorite computer algebra system)

$f = 3xy - 4z$

$g = 3x + y - 2xz$

Solve $[\{D[f,x] == \lambda D[g,x], D[f,y] == \lambda D[g,y], D[f,z] == \lambda D[g,z], 3x + y - 2x\,z == 1\}]$
The solutions are

- $\lambda = \sqrt{6}, (x,y,z) = (\sqrt{2/3}, 1/2, (12 - \sqrt{6})/8)$ and
- $\lambda = -\sqrt{6}, (x,y,z) = (-\sqrt{2/3}, 1/2, (12 + \sqrt{6})/8)$.

17. Many solutions are returned by *Mathematica.* They are

- $(0, -1, 0)$ for $\lambda = -3/2$
- $(0, 1, 0)$ for $\lambda = 3/2$
- $(-2/3, -2/3, -1/3)$ and $(2/3, -2/3, 1/3)$ for $\lambda = -4/3$
- $(-1, 0, 0)$ and $(1, 0, 0)$ for $\lambda = -1$
- $(0, 0, -1)$ and $(0, 0, 1)$ for $\lambda = 0$ and
- $(\sqrt{11/2}/8, -3/8, -3\sqrt{11/2}/8)$ and $(-\sqrt{11/2}/8, -3/8, 3\sqrt{11/2}/8)$ for $\lambda = 1/8$.

21. The symmetry of the problem suggests the answer, but we are maximizing $f(x,y,z) = xyz$ subject to the constraint $g(x,y,z) = x + y + z = 18$. We solve the system

$$\begin{cases} yz = \lambda \\ xz = \lambda \\ xy = \lambda \\ x + y + z = 18. \end{cases}$$

None of the solutions that corresponds to one of x, y, and z being zero is a maximum. The solution we get is $x = y = z$, so $3x = 18$, so the maximum product occurs at the point $(6, 6, 6)$.

25. We are maximizing $f(r,h) = \pi r^2 h$ subject to the constraint that $g(r,h) = 2\pi rh + 2\pi r^2 = c$. We solve the system

$$\begin{cases} 2\pi rh = \lambda(2\pi h + 4\pi r) \\ \pi r^2 = 2\lambda\pi r \\ 2\pi rh + 2\pi r^2 = c. \end{cases}$$

Since $r \neq 0$ the second equation implies that $r = 2\lambda$, so, substituting this into the first equation, we see that $h = 2r$. Hence, the height should equal the diameter.

29. A sphere centered at the origin has equation $x^2+y^2+z^2 = r^2$. Thus we want to maximize $f(x,y,z) = x^2+y^2+z^2$ subject to the constraint $g(x,y,z) = 3x^2+2y^2+z^2 = 6$. We can solve this using Lagrange multipliers, but we must make sure we find an *inscribed* sphere. We consider the system

$$\begin{cases} 2x = 6\lambda x & 1^{\text{st}} \text{ equation gives } x = 0 \text{ or } \lambda = 1/3 \\ 2y = 4\lambda y & 2^{\text{nd}} \text{ equation gives } y = 0 \text{ or } \lambda = 1/2 \\ 2z = 2\lambda z & 3^{\text{rd}} \text{ equation gives } z = 0 \text{ or } \lambda = 1 \\ 3x^2+2y^2+z^2 = 6 & \text{(Note that we can't have } x = y = z = 0 \\ & \quad \text{and still satisfy the constraint.)} \end{cases}$$

Thus if $\lambda = 1/3$, $y = z = 0$ and the constraint implies $x = \pm\sqrt{2}$. If $\lambda = 1/2$, $x = z = 0$ and $y = \pm\sqrt{3}$. Finally, if $\lambda = 1$, then $x = y = 0$ and $z = \pm\sqrt{6}$. Comparing values, we have

$$f(\pm\sqrt{2},0,0) = 2, \quad f(0,\pm\sqrt{3},0) = 3, \quad f(0,0,\pm\sqrt{6}) = 6,$$

so that it's tempting to say that the largest sphere has a radius of $\sqrt{6}$. However, such a sphere is not actually inscribed in the ellipsoid. The largest sphere that actually remains inscribed in the ellipsoid has a radius of $\sqrt{2}$.

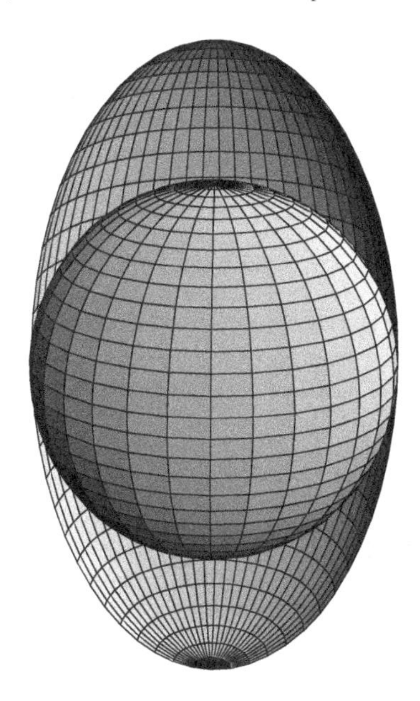

31. This is just Exercise 22 translated by $(2,5,-1)$. We are minimizing $f(x,y,z) = (x-2)^2+(y-5)^2+(z+1)^2$ with the constraints $g_1(x,y,z) = x-2y+3z = 8$ and $g_2(x,y,z) = 2z-y = 3$. We solve the system

$$\begin{cases} 2(x-2) = \lambda \\ 2(y-5) = -2\lambda - \mu \\ 2(z+1) = 3\lambda + 2\mu \\ x - 2y + 3z = 8 \\ 2z - y = 3. \end{cases}$$

Eliminate μ by combining the second and third equations and then substitute $2(x-2)$ for λ. Solve to get a critical point at (9/2, 2, 5/2).

35. **(a)** $f(x, y) = x + y$ with the constraint $xy = 6$ so we solve the system

$$\begin{cases} 1 = 2\lambda y \\ 1 = 2\lambda x \\ xy = 6. \end{cases}$$

So $x = y$ and the critical points are at $\pm(\sqrt{6}, \sqrt{6})$.

(b) The constraint curve is not connected. There are two distinct components. Although $(-\sqrt{6}, -\sqrt{6})$ produces a local maximum of $-2\sqrt{6}$ on its component, the value of the function at any point on the other component is greater. Similarly, $(\sqrt{6}, \sqrt{6})$ produces a local minimum of $2\sqrt{6}$ on its component, but the value of the function at any point on the other component is less.

39. We want to find the extreme values of the function $f(x, y) = x^2 + y^2$ (the square of the distance from the point (x, y) to the origin) subject to the constraint $g(x, y) = 3x^2 - 4xy + 3y^2 = 50$. (Note that there will be a global maximum and a global minimum by the extreme value theorem since the ellipse is a compact set in $\mathbf{R}^2$.) We solve the system

$$\begin{cases} 2x = \lambda(6x - 4y) \\ 2y = \lambda(-4x + 6y) \\ 3x^2 - 4xy + 3y^2 = 50. \end{cases}$$

The first two equations together imply

$$\frac{1}{\lambda} = \frac{6x - 4y}{2x} = \frac{-4x + 6y}{2y} \iff 3 - \frac{2y}{x} = 3 - \frac{2x}{y} \iff y^2 = x^2.$$

Thus $y = \pm x$. If $y = x$, then the last equation becomes

$$3x^2 - 4x^2 + 3x^2 = 50 \iff x^2 = 25 \iff x = \pm 5.$$

Thus there are two critical points $(5, 5)$ and $(-5, -5)$. If $y = -x$, then the last equation becomes

$$3x^2 + 4x^2 + 3x^2 = 50 \iff x^2 = 5 \iff x = \pm\sqrt{5}.$$

Hence there are two more critical points $(\sqrt{5}, -\sqrt{5})$ and $(-\sqrt{5}, \sqrt{5})$. Finally, we have

$$f(5, 5) = f(-5, -5) = 50 \quad \text{and} \quad f(\sqrt{5}, -\sqrt{5}) = f(-\sqrt{5}, \sqrt{5}) = 10,$$

so that $(5, 5)$ and $(-5, -5)$ are the points on the ellipse farthest from the origin and $(\sqrt{5}, -\sqrt{5})$ and $(-\sqrt{5}, \sqrt{5})$ are the points nearest the origin.

43. **(a)** In order for $(\boldsymbol{\lambda}, \mathbf{a})$ to be a solution of the constrained problem, $(\boldsymbol{\lambda}, \mathbf{a})$ must solve the system

$$\begin{cases} f_{x_i}(\mathbf{a}) = \sum_{j=1}^{k} \lambda_j (g_j)_{x_i}(\mathbf{a}) & \text{for } 1 \le i \le n \\ g_j(\mathbf{a}) = c_j & \text{for } 1 \le j \le k. \end{cases}$$

On the other hand, an unconstrained critical point for L must be where all first partials are zero. In other words, we must have

$$L_{l_j} = 0, \quad 1 \le j \le k \quad \text{and} \quad L_{x_j} = 0, \quad 1 \le j \le n.$$

Upon explicit calculation of the partials these equations are:

$$\begin{cases} f_{l_j}(\mathbf{a}) - (g_j(\mathbf{a}) - c_j) = 0 & \text{for } 1 \le j \le k, \text{ and} \\ f_{x_j}(\mathbf{a}) - \sum_{i=1}^{k} \lambda_i (g_i)_{x_j}(\mathbf{a}) = 0 & \text{for } 1 \le j \le n. \end{cases}$$

This is the same system as that for the constrained case.

(b) Calculate the Hessian in four blocks. All of the entries in the upper left $k \times k$ block are 0. This is because the entry in position (i, j) is $L_{l_i l_j}$ and the highest power of any l_i appearing in L is 1. The top right block with k rows and n columns gives back the negative first partials of the constraint conditions because the entry in position $(k + i, j)$ is $L_{x_i l_j} = -(g_j - c_j)_{x_i} = -(g_j)_{x_i}$. The lower left block of n rows and k columns is just the transpose of this last block. The lower right $n \times n$ block is such that the entry in position $(k + i, k + j) = L_{x_i x_j} = (f - \sum_{q=1}^{k} l_q g_q)_{x_i x_j}$. When $\boldsymbol{\lambda}$ and $\mathbf{a}$ are substituted for $\mathbf{l}$ and $\mathbf{x}$, the desired matrix is obtained.

4.4 Some Applications of Extrema

3. (a) As in the text, the function $D(a, b)$ will be the sum of the squares of the differences between the observed y values and the y values on the curve $y = a/x + b$. This means that

$$D(a,b) = \sum_{i=1}^{n} (y_i - (a/x_i + b))^2.$$

(b) Make the substitution $X_i = 1/x_i$ and then fit the line $y = aX + b$ to this transformed data using Proposition 4.1. We get

$$a = \frac{n\sum X_i y_i - \left(\sum X_i\right)\left(\sum y_i\right)}{n\sum X_i^2 - \left(\sum X_i\right)^2} \quad \text{and} \quad b = \frac{\left(\sum X_i^2\right)\left(\sum y_i\right) - \left(\sum X_i\right)\left(\sum X_i y_i\right)}{n\sum X_i^2 - \left(\sum X_i\right)^2}.$$

Transform the data back, replacing X_i with $1/x_i$, then the curve of the form $y = a/x + b$ that best fits the data has

$$a = \frac{n\sum y_i/x_i - \left(\sum 1/x_i\right)\left(\sum y_i\right)}{n\sum 1/x_i^2 - \left(\sum 1/x_i\right)^2} \quad \text{and} \quad b = \frac{\left(\sum 1/x_i^2\right)\left(\sum y_i\right) - \left(\sum 1/x_i\right)\left(\sum y_i/x_i\right)}{n\sum 1/x_i^2 - \left(\sum 1/x_i\right)^2}.$$

7. (a) We are required to show that $\mathbf{F}$ is a gradient (conservative) vector field. Clearly if $V(x, y) = x^2 + 2xy + 3y^2 + x + 2y$ then $-\nabla V = (-2x - 2y - 1)\mathbf{i} + (-2x - 6y - 2)\mathbf{j} = \mathbf{F}$.

(b) We find equilibrium points of $\mathbf{F}$ when $\mathbf{F} = \mathbf{0}$. Solve the system of equations

$$\begin{cases} -2x - 2y = 1 \\ -2x - 6y = 2 \end{cases}$$

and find one solution at $(-1/4, -1/4)$. The Hessian is

$$HV = \begin{bmatrix} 2 & 2 \\ 2 & 6 \end{bmatrix}$$

so both d_1 and d_2 are positive so the equilibrium is stable.

11. Maximize $R(x, y, z) = xyz^2 - 25000x - 25000y - 25000z$ subject to the constraint $x + y + z = 200000$. Our system of equations is

$$\begin{cases} yz^2 - 25000 = \lambda \\ xz^2 - 25000 = \lambda \\ 2xyz - 25000 = \lambda \\ x + y + z = 200000. \end{cases}$$

The hidden condition is that all of the variables are non-negative. This means that we are finding a maximum on the triangular portion of the plane that lies in the first octant. The maximum revenue will occur at a boundary point or at a critical point. Along the boundary at least one of the variables is 0 and the revenue is at most 0 when at least one of x, y and z is 0. We will see the value of R at the critical point is greater and therefore that it is our global maximum. Assume none of the variables is zero. Then, from the first two equations, since $z \neq 0$ then $x = y$. From the third equation paired with either of the first two we see that $z = 2x = 2y$. Finally, since their sum is 200000 we find the solution (50000, 50000, 100000) is where the maximum revenue occurs.

15. (a) This is an example of the Cobb-Douglas production function with $p = w = 1$ (see Example 5 from the text). The only critical point will be $(K, L) = ((1/3)360000, (2/3)360000) = (120000, 240000)$.

(b) $\partial Q/\partial K = 20(L/K)^{2/3}$ and so at (120000, 240000), $\partial Q/\partial K = 20(2)^{2/3}$. On the other hand, $\partial Q/\partial L = 40(K/L)^{1/3}$ and so at (120000, 240000), $\partial Q/\partial L = 40(1/2)^{1/3}$. These quantities are equal at the critical point.

True/False Exercises for Chapter 4

1. True.
3. True.
5. True.
7. False. (f is most sensitive to changes in y.)

9. False.
11. True.
13. False. (Consider the function $f(x, y) = x^2 + y^2$.)
15. True.
17. False. (The point is not a critical point of the function.)
19. True.
21. False. (The critical point is a saddle point.)
23. False. (Extrema may also occur at points where $g = c$ and $\nabla g = \mathbf{0}$.)
25. False. (You will have to solve a system of 7 equations in 7 unknowns.)
27. True.
29. False. (The equilibrium points are the critical points of the potential function.)

Miscellaneous Exercises for Chapter 4

3. We are asked to maximize the profit $P(x, y) = (x - 2)(80 - 100x + 40y) + (y - 4)(20 + 60x - 35y) = -100x^2 + 40x - 35y^2 + 80y + 100xy - 240$. The partial derivatives are $P_x(x, y) = -200x + 100y + 40$ and $P_y(x, y) = 100x - 70y + 80$. These are both zero at (27/10, 5). You can read the Hessian right off the first derivatives and you see that $d_1 = -200 < 0$ and $d_2 = 4000 > 0$ so profit is maximized when you charge \$2.70 for Mocha and \$5 for Kona.

7. (a) $f_x(x, y) = 2x(-3y + 4x^2)$ while $f_y(x, y) = 2y - 3x^2$. From f_x we see that either $x = 0$ or $y = (4/3)x^2$. But from the second equation $y = (3/2)x^2$. So we conclude that the only solution is at (0, 0).

(b) $f_{xx}(x, y) = 6(-y + 4x^2), f_{xy}(x, y) = -6x$, and $f_{yy}(x, y) = 2$. At the origin, the Hessian is $\begin{bmatrix} 0 & 0 \\ 0 & 2 \end{bmatrix}$ and so the determinant is 0 and the critical point is degenerate.

(c) If $y = mx$ then the original equation becomes $F(x) = m^2x^2 - 3mx^3 + 2x^4$. We calculate $F'(x) = 2m^2x - 9mx^2 + 8x^3 = 2x(m^2 - 9mx/2 + 4x^2)$. From the second derivative we see that $F''(x) = 2m^2 - 18mx + 24x^2$. This is positive at $x = 0$ for all $m \neq 0$ so there is a minimum for $x = 0$ along any line other than the two axes. When $m = 0, F'(x) = 8x^3$ and so the first derivative test implies that there is a minimum at $x = 0$ when $m = 0$. Finally, consider $G(y) = f(0, y) = y^2$. This clearly has a minimum at $y = 0$. We've shown that along any line through the origin, f has a minimum at (0, 0).

(d) Consider $g(x) = f(x, 3x^2/2) = (-x^2/2)(x^2/2) = -x^4/4$. From the derivative $g'(x) = -x^3$ we see that g has a maximum at $x = 0$ and hence f has a maximum at the origin when constrained to the given parabola. This means that the origin is actually a saddle point for f.

(e) A portion of the surface is shown below.

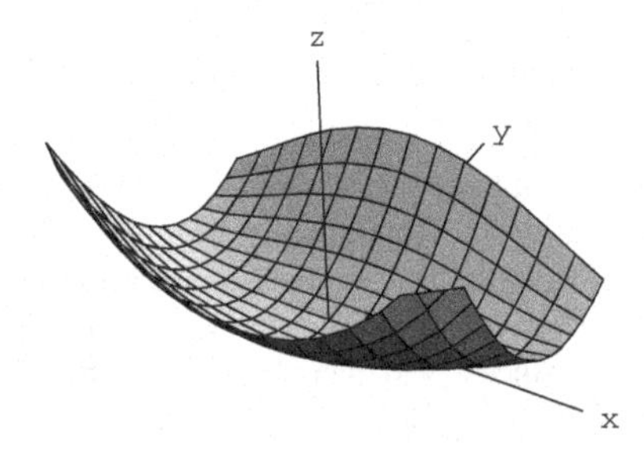

9. (a) Here we are finding the critical points of $f(x, y, z) = xy$ subject to the constraint $g(x, y, z) = x^2 + y^2 + z^2 - 1 = 0$. So taking the partials of $f(x, y, z) = \lambda g(x, y, z)$ along with the constraint we get the following system of equations.

$$\begin{cases} y = 2\lambda x \\ x = 2\lambda y \\ 0 = 2\lambda z \\ 1 = x^2 + y^2 + z^2 \end{cases}$$

This problem is very similar to Exercise 8 and so it is no surprise that we again get four critical points corresponding to $\lambda = \pm 1/2$. They are $(1/\sqrt{2}, 1/\sqrt{2}, 0), (-1/\sqrt{2}, 1/\sqrt{2}, 0), (1/\sqrt{2}, -1/\sqrt{2}, 0)$, and $(-1/\sqrt{2}, -1/\sqrt{2}, 0)$. We also get critical points at the two poles corresponding to $\lambda = 0$. These are at $(0, 0, \pm 1)$.

(b) Of course, it is harder to represent this situation than its lower-dimensional counterpart. Here are some level sets, the unit sphere and the critical points.

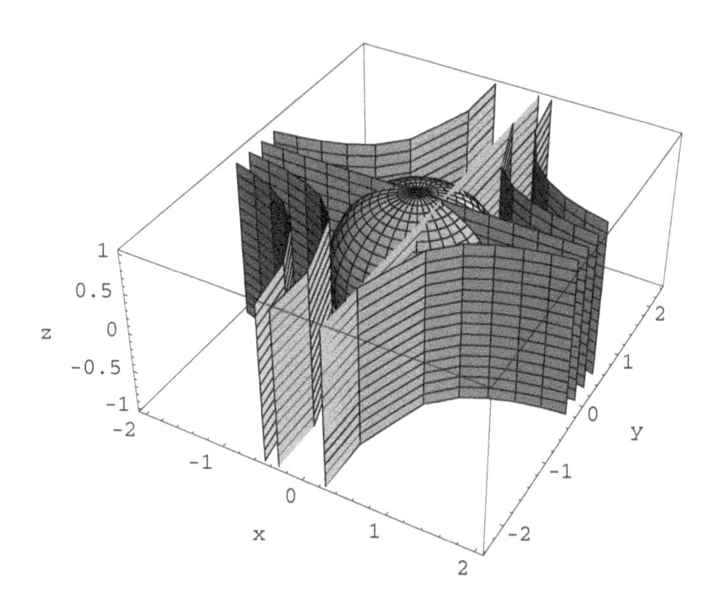

(c) The arguments that f has a constrained max at $\pm(1/\sqrt{2}, 1/\sqrt{2}, 0)$ and has a constrained min at $\pm(1/\sqrt{2}, -1/\sqrt{2}, 0)$ are the same as in Exercise 8. The two poles must be saddle points. If you travel in a direction where $y = x, f(x, y)$ is increasing while if you travel in a direction where $y = -x, f(x, y)$ is decreasing. So there are saddle points at $(0, 0, \pm 1)$.

15. Minimize $M(x, y, z) = x^2 + y^2 + z^2$ subject to $x^2 - (y - z)^2 = 1$. We solve

$$\begin{cases} 2x = 2\lambda x \\ 2y = -2\lambda(y - z) \\ 2z = 2\lambda(y - z) \\ x^2 - (y - z)^2 = 1. \end{cases}$$

Since the last equation implies that $x \neq 0$, the first equation gives us that $\lambda = 1$, so $y = z = 0$ and thus $x = \pm 1$. The minimum distance is, therefore, 1.

19. The two equations are $x = y/2 - 1$ and $x = y^2$. We will minimize the square of the distance between a point (x_1, y_1) on the line and a point (x_2, y_2) on the parabola. Maximize $f(y_1, y_2) = (y_1/2 - 1 - y_2^2)^2 + (y_2 - y_1)^2$. Take the first partials:

$$f_{y_1}(y_1, y_2) = \frac{5}{2}y_1 - 1 - y_2^2 - 2y_2 \quad \text{and}$$
$$f_{y_2}(y_1, y_2) = 4y_2^3 - 2y_1y_2 + 6y_2 - 2y_1.$$

Set these equal to zero and solve to find the critical point at $(y_1, y_2) = (5/8, 1/4)$. The minimal distance is therefore $3\sqrt{5}/8$.

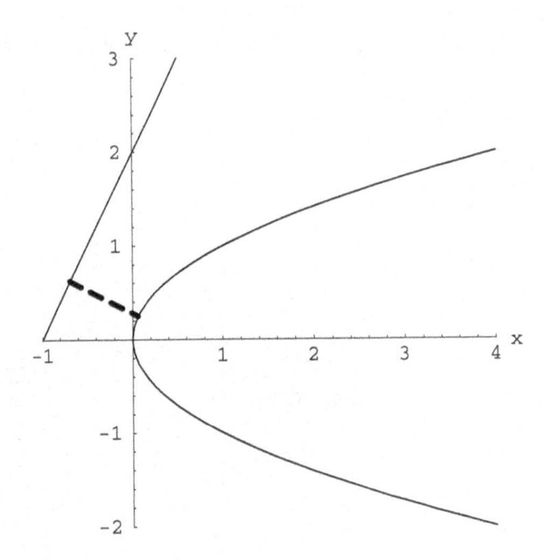

21. We are minimizing the square of the distance $f(x, y) = (x - x_0)^2 + (y - y_0)^2$ subject to the constraint $ax + by = d$. We solve

$$\begin{cases} 2(x - x_0) = a\lambda \\ 2(y - y_0) = b\lambda \\ ax + by = d. \end{cases}$$

Solving, we see that $x = (a\lambda + 2x_0)/2$ and $y = (b\lambda + 2y_0)/2$ so substituting for x and y in the third equation $(a^2 + b^2)\lambda = 2(d - ax_0 - by_0)$. Also substituting for x and y in f we see that

$$f(x, y) = \left(\frac{a\lambda}{2}\right)^2 + \left(\frac{b\lambda}{2}\right)^2 = \left(\frac{a^2 + b^2}{4}\right)\lambda^2 = \frac{a^2 + b^2}{4}\left(\frac{2(d - ax_0 - by_0)}{a^2 + b^2}\right)^2$$
$$= \frac{(d - ax_0 - by_0)^2}{a^2 + b^2}$$

so the distance D is the square root of this: $D = \dfrac{|ax_0 + by_0 - d|}{\sqrt{a^2 + b^2}}$.

25. (a) To set things up using Lagrange multipliers, we solve

$$\begin{cases} 2ax + 2by = 2\lambda x \\ 2bx + 2cy = 2\lambda y \\ x^2 + y^2 = 1 \end{cases} \Leftrightarrow \begin{cases} (a - \lambda)x + by = 0 \\ bx + (c - \lambda)y = 0 \\ x^2 + y^2 = 1. \end{cases}$$

In the last system, multiply the first equation by $\lambda - c$ and the second by b, then add to obtain:

$$((a - \lambda)(\lambda - c) + b^2)x = 0.$$

Now multiply the first equation by b and the second by $\lambda - a$, then add to get:

$$(b^2 + (\lambda - a)(c - \lambda))y = 0.$$

Since $x^2 + y^2 = 1$, we cannot have both x and y equal to 0. Thus

$$b^2 + (\lambda - a)(c - \lambda) = 0 \Leftrightarrow \lambda^2 - (a + c)\lambda + ac - b^2 = 0.$$

Hence

$$\lambda_1, \lambda_2 = \frac{(a + c) \pm \sqrt{(a + c)^2 - 4(ac - b^2)}}{2}.$$

(b) Rewriting, $\lambda_1, \lambda_2 = \dfrac{(a + c) \pm \sqrt{(a - c)^2 + 4b^2}}{2}$. $(a - c)^2 + 4b^2 \geq 0$ so the eigenvalues are always real.

Chapter 5

Multiple Integration

5.1 Introduction: Areas and Volumes

3.

$$\int_{-2}^{4}\int_{0}^{1}(xe^{y})\,dy\,dx = \int_{-2}^{4}(xe^{y})\Big|_{y=0}^{y=1}dx = \int_{-2}^{4}(x(e-1))\,dx$$
$$= \frac{x^2}{2}(e-1)\Big|_{-2}^{4} = (8-2)(e-1) = 6(e-1).$$

7. (a) Here we are fixing x and finding the area of the slices:

$$A(x) = \int_{0}^{2}(x^2+y^2+2)\,dy = \left(x^2y+\frac{y^3}{3}+2y\right)\Big|_{0}^{2} = 2x^2+20/3.$$

Now we "add up the areas of these slices":

$$V = \int_{-1}^{2}A(x)\,dx = \int_{-1}^{2}(2x^2+20/3)\,dx = \left(\frac{2}{3}x^3+\frac{20}{3}x\right)\Big|_{-1}^{2} = \left(\frac{16}{3}+\frac{40}{3}\right)-\left(-\frac{2}{3}-\frac{20}{3}\right) = 26.$$

(b) Now we fix y and find the area of the slices:

$$A(y) = \int_{-1}^{2}(x^2+y^2+2)\,dx = \left(\frac{x^3}{3}+y^2x+2x\right)\Big|_{-1}^{2}$$
$$= \left(\frac{8}{3}+2y^2+4\right)-\left(-\frac{1}{3}-y^2-2\right) = 9+3y^2.$$

Adding up the area of these slices:

$$V = \int_{0}^{2}A(y)\,dy = \int_{0}^{2}(9+3y^2)\,dy = (9y+y^3)\Big|_{0}^{2} = 26.$$

11. This is the volume of the region bounded by the paraboloid $z = 16-x^2-z^2$, the xy-plane, and the planes $x = 1$, $x = 3$, $y = -2$, and $y = 2$. The volume is

$$V = \int_{1}^{3}\int_{-2}^{2}(16-x^2-y^2)\,dy\,dx = \int_{1}^{3}\left(16y-x^2y-\frac{y^3}{3}\right)\Big|_{-2}^{2}dx = \int_{1}^{3}\left(64-4x^2-\frac{16}{3}\right)dx$$
$$= \left(64x-\frac{4}{3}x^3-\frac{16}{3}x\right)\Big|_{1}^{3} = (192-36-16)-(64-4/3-16/3) = 248/3.$$

15.

$$\int_{-5}^{5}\int_{-1}^{2}(5-|y|)\,dx\,dy = \int_{-5}^{5}(5-|y|)x\Big|_{x=-1}^{2}dy$$
$$= \int_{-5}^{5}(5-|y|)\cdot 3\,dy = 150-3\int_{-5}^{5}|y|\,dy$$
$$= 150-3\int_{-5}^{0}(-y)\,dy-3\int_{0}^{5}y\,dy$$
$$= 150+\frac{3}{2}y^2\Big|_{-5}^{0}-\frac{3}{2}y^2\Big|_{0}^{5} = 150-\frac{75}{2}-\frac{75}{2} = 75.$$

The iterated integral gives the volume of the region bounded by the graph of $z = 5 - |y|$, the xy-plane, and the planes $x = -1$, $x = 2$, $y = -5$, $y = 5$. (The solid so described is a rectangular prism.)

5.2 Double Integrals

1. Since the integrand $f(x, y) = y^3 + \sin 2y$ is continuous, the double integral $\iint_R (y^3 + \sin 2y)\, dA$ exists by Theorem 2.4. Now consider a Riemann sum corresponding to the double integral that we obtain by partitioning the rectangle $[0, 3] \times [-1, 1]$ symmetrically with respect to the x-axis and by choosing test points $\mathbf{c}_{ij}$ in each subrectangle that are also symmetric with respect to the x-axis. Then

$$S = \sum_{i,j} f(\mathbf{c}_{ij})\, \Delta A_{ij} = \sum_{i,j} (y_{ij}^3 + \sin 2y_{ij})\, \Delta A_{ij}$$

(where y_{ij} denotes the y-coordinate of $\mathbf{c}_{ij}$) must be zero since the terms cancel in pairs because $f(x, -y) = -f(x, y)$. When we shrink the rectangles in the limit, we can arrange to preserve all the symmetry. Hence the limit under such restrictions must be zero and thus the overall limit (which must exist in view of Theorem 2.4) must also be zero.

5. $\displaystyle\int_0^2 \int_0^{y^2} y\, dx\, dy = \int_0^2 xy\Big|_0^{y^2} dy = \int_0^2 y^3\, dy = \frac{y^4}{4}\Big|_0^2 = 4$. The region over which we are integrating is:

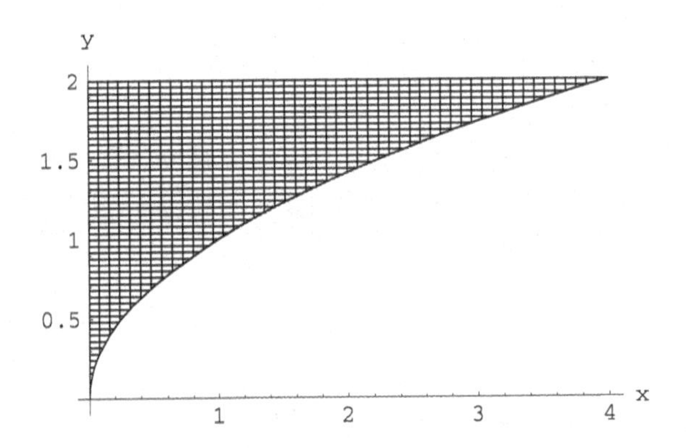

11. $\displaystyle\int_0^1 \int_{-\sqrt{1-x^2}}^{\sqrt{1-x^2}} 3\, dy\, dx = \int_0^1 3y\Big|_{-\sqrt{1-x^2}}^{\sqrt{1-x^2}} dx = \int_0^1 6\sqrt{1 - x^2}\, dx =$ (using the substitution $x = \sin t$) $= 3\pi/2$. You can also see that the region over which we are integrating is a half-circle of radius 1 so we have found the volume of the cylinder over this region of height 3. This figure is:

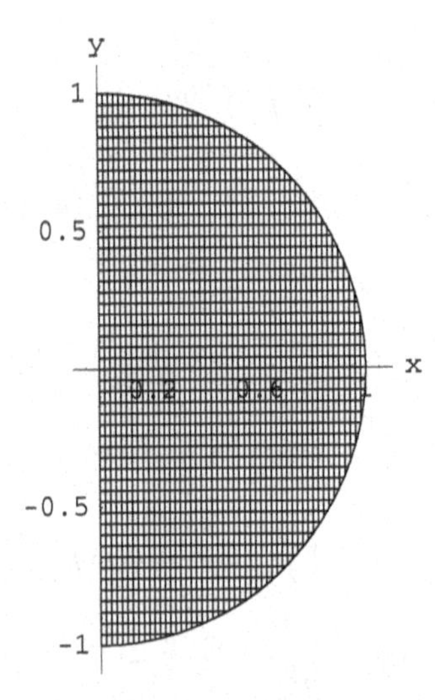

15. A quick sketch of the region over which we are integrating helps us set up our double integral.

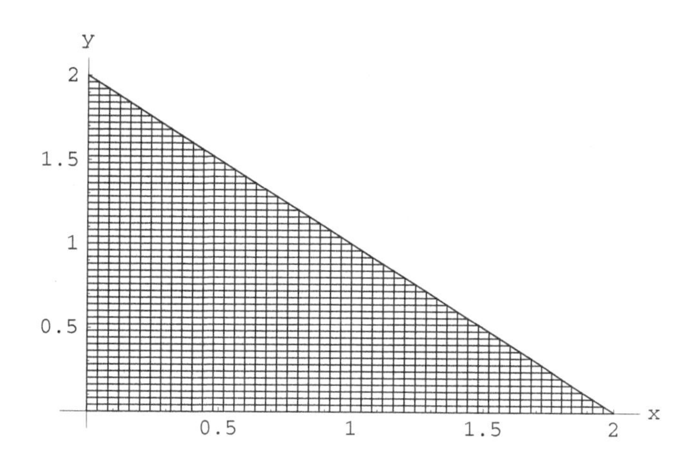

$$\int_0^2 \int_0^{2-x} (1 - xy)\, dy\, dx = \int_0^2 \left(y - \frac{xy^2}{2} \right) \Big|_0^{2-x} dx = \int_0^2 (2 - 3x + 2x^2 - x^3/2)\, dx$$
$$= \left(2x - \frac{3}{2}x^2 + \frac{2}{3}x^3 - \frac{x^4}{8} \right) \Big|_0^2 = 4 - 6 + 16/3 - 2 = 4/3.$$

17. We can easily determine the limits of integration from the sketch and/or by solving for where $x + y = 2$ intersects the parabola $y^2 - 2y - x = 0$.

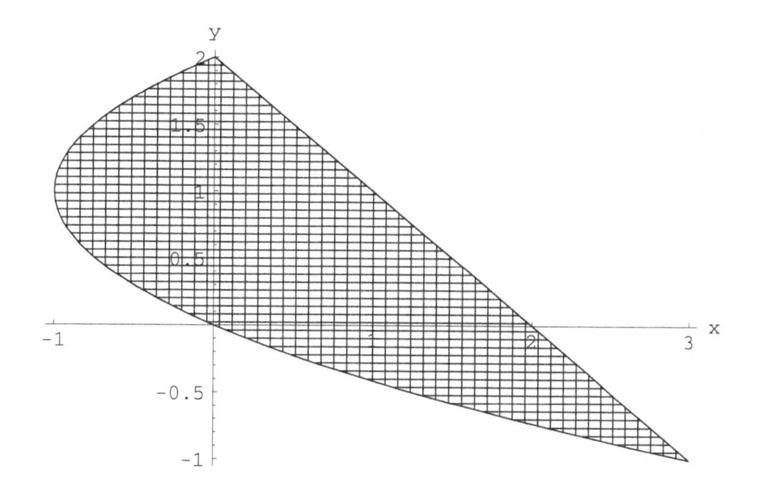

$$\int_{-1}^2 \int_{y^2-2y}^{2-y} (x + y)\, dx\, dy = \int_{-1}^2 \left(\frac{x^2}{2} + xy \right) \Big|_{y^2-2y}^{2-y} dy = \int_{-1}^2 \left(-\frac{y^4}{2} + y^3 - \frac{y^2}{2} + 2 \right) dy$$
$$= \left(-\frac{y^5}{10} + \frac{y^4}{4} - \frac{y^3}{6} + 2y \right) \Big|_{-1}^2 = \frac{99}{20}.$$

23. As in the proof of property 1 in the text, we note that the Riemann sum whose limit is

$$\iint_R cf\, dA \text{ is } \sum_{i,j=1}^n cf(\mathbf{c}_{ij})\Delta A_{ij} = c \sum_{i,j=1}^n f(\mathbf{c}_{ij})\Delta A_{ij} \to c \iint_R f\, dA.$$

25. Define $f^+ = \max(f, 0)$ and $f^- = \max(-f, 0)$. Note that both f^+ and f^- have only non-negative values. Then $f = f^+ - f^-$ and $|f| = f^+ + f^-$. Since $f^{\pm} \leq |f| = f^+ + f^-$ we can see that $|f|$ is Riemann integrable. Also we can use property 2 to conclude that

$$\left|\iint_R f\,dA\right| = \left|\iint_R (f^+ - f^-)\,dA\right| = \left|\iint_R f^+\,dA - \iint_R f^-\,dA\right|$$
$$\leq \iint_R f^+\,dA + \iint_R f^-\,dA = \iint_R |f|\,dA.$$

29. We integrate $\displaystyle\int_{-a}^{a}\int_{-\sqrt{b^2-b^2x^2/a^2}}^{\sqrt{b^2-b^2x^2/a^2}} dy\,dx = 2\int_{-a}^{a}\sqrt{b^2 - \frac{b^2x^2}{a^2}}\,dx = \frac{2b}{a}\left(\int_{-a}^{a}\sqrt{a^2-x^2}\,dx\right) = \frac{b}{a}(\pi a^2) = \pi ab.$

33. First, note that the integrand is continuous; hence the integral as the limit of Riemann sums must exist. Second, note that the region D is symmetric with respect to the x-axis. Next, note that we can break up the integral as

$$\iint_D (y^3 + e^{x^2}\sin y + 2)\,dA = \iint_D y^3\,dA + \iint_D e^{x^2}\sin y\,dA + \iint_D 2\,dA.$$

Consider first $\iint_D y^3\,dA$ and note that the integrand, y^3, is an odd function. Hence, in a Riemann sum, we can arrange to partition any rectangle that contains D in such a way that for every subrectangle above the x-axis (i.e., where $y > 0$), there is a corresponding "mirror image" subrectangle—with the same area—*below* the x-axis (where $y < 0$). Then the "test points" in each pair of subrectangles may be chosen to have *opposite* y-coordinates. (See the figure below.)

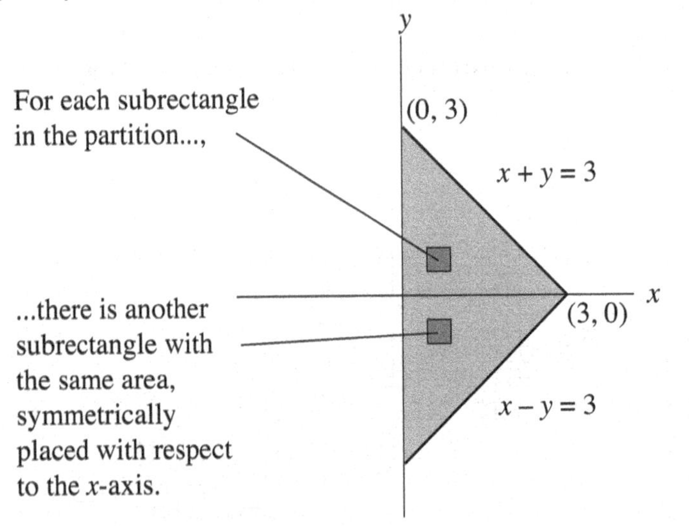

The Riemann sum corresponding to this partition will be

$$\sum_{i,j} y_{ij}^3\,\Delta A_{ij} = 0,$$

since the terms of the sum will cancel in pairs. Thus, even when we take the limit of this sum as $\Delta A_{ij} \to 0$, we still obtain zero. Therefore, we conclude that $\iint_D y^3\,dA = 0$. Using a similar argument, we find that $\iint_D e^{x^2}\sin y\,dA = 0$ as well. Hence

$$\iint_D (y^3 + e^{x^2}\sin y + 2)\,dA = \iint_D y^3\,dA + \iint_D e^{x^2}\sin y\,dA + \iint_D 2\,dA$$
$$= 0 + 0 + 2\iint_D dA = 2(\text{area of } D)$$
$$= 2(9) = 18.$$

39. By symmetry we see that the volume is four times the volume of the piece over the first quadrant $(x, y \geq 0)$. In this region $|x| = x$ and $|y| = y$ so the volume is

$$4\int_0^2 \int_0^{2-x} (2 - x - y)\, dy\, dx = 4\int_0^2 (2y - xy - y^2/2)\Big|_0^{2-x} dx = 4\int_0^2 (2 - 2x + x^2/2)\, dx$$
$$= 4(2x - x^2 + x^3/6)\Big|_0^2 = 16/3.$$

41. **(a)** If x is rational, then $\int_0^2 f(x, y)\, dy = \int_0^2 1\, dy = 2$. If x is irrational, then $\int_0^2 f(x, y)\, dy = \int_0^1 0\, dy + \int_1^2 2\, dy = 2$.

(b) Using our answer from part (a), $\int_0^1 \int_0^2 f(x, y)\, dy\, dx = \int_0^1 2\, dx = 2$.

(c) If $\mathbf{c}_{ij}$ has a rational x coordinate, then $f(\mathbf{c}_{ij}) = 1$ and so the Riemann sum will converge to the area of the region, which is 2.

(d) In this case $f(\mathbf{c}_{ij}) = 1$ for our points in the region $[0, 1] \times [0, 1]$ and $f(\mathbf{c}_{ij}) = 2$ for our points in the region $[0, 1] \times [1, 2]$. In short, the Riemann sums will converge to $(1)(1) + (2)(1) = 3$.

(e) As we saw in parts (c) and (d), the Riemann sum does not have a well defined limit and so f fails to be integrable on R, even though in part (b) we actually computed the iterated integral.

5.3 Changing The Order of Integration

This is a good section to explore with a computer system.

1. **(a)**

$$\int_0^2 \int_{x^2}^{2x} (2x + 1)\, dy\, dx = \int_0^2 (2x + 1)(2x - x^2)\, dx = \int_0^2 (-2x^3 + 3x^2 + 2x)\, dx$$
$$= \left(-\frac{x^4}{2} + x^3 + x^2\right)\Big|_0^2 = 4.$$

(b) The region of integration is bounded above by $y = 2x$ and below by $y = x^2$:

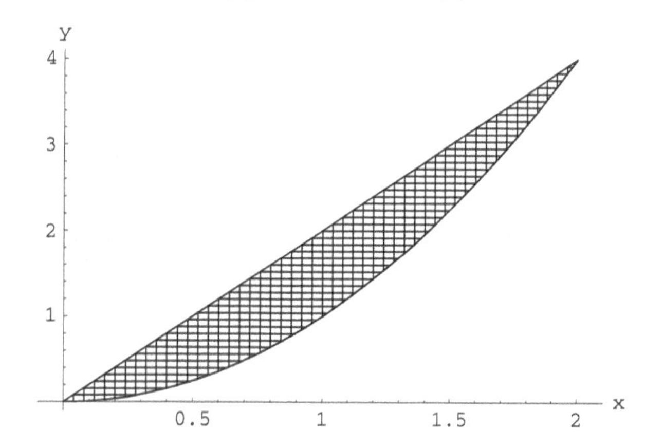

(c)

$$\int_0^4 \int_{y/2}^{\sqrt{y}} (2x + 1)\, dx\, dy = \int_0^4 (x^2 + x)\Big|_{y/2}^{\sqrt{y}} dy = \int_0^4 \left(-\frac{y^2}{4} + \frac{y}{2} + \sqrt{y}\right) dy$$
$$= \left(-\frac{y^3}{12} + \frac{y^2}{4} + \frac{2y^{3/2}}{3}\right)\Big|_0^4 = 4.$$

5. The region of integration is:

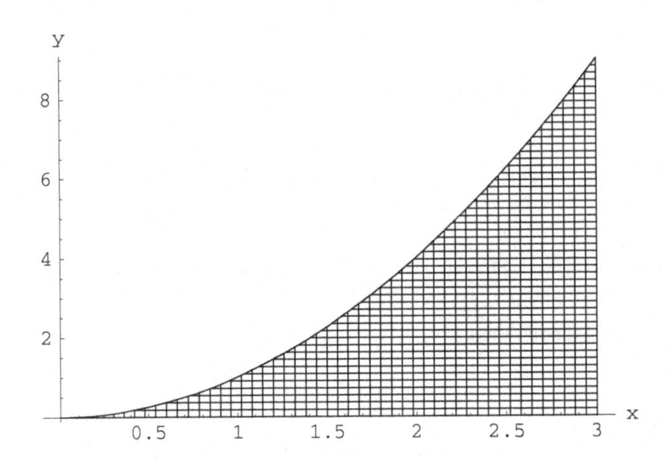

$$\int_0^9 \int_{\sqrt{y}}^3 (x+y)\,dx\,dy = \int_0^9 \frac{1}{2}(-2y^{3/2}+5y+9)\,dy = 891/20 \quad \text{and}$$

$$\int_0^3 \int_0^{x^2} (x+y)\,dy\,dx = \int_0^9 (x^4/2+x^3)\,dx = 891/20.$$

9. The region of integration is:

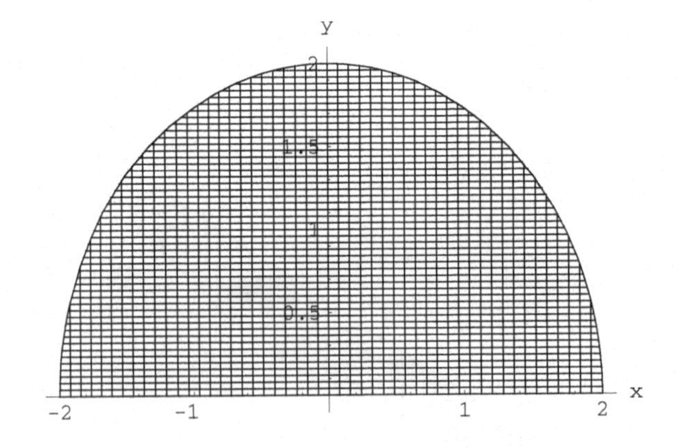

$$\int_0^2 \int_{-\sqrt{4-y^2}}^{\sqrt{4-y^2}} y\,dx\,dy = \int_0^2 (2y\sqrt{4-y^2})\,dy = 16/3 \quad \text{and}$$

$$\int_{-2}^2 \int_0^{\sqrt{4-x^2}} y\,dy\,dx = \int_{-2}^2 (-x^2/2+2)\,dx = 16/3.$$

13. The limits of integration of the first integral describe the region D_1 bounded on the left by the x-axis, on the right by $x = \sqrt{y/3}$ (or, equivalently, by $y = 3x^2$) and on top by $x = 8$.

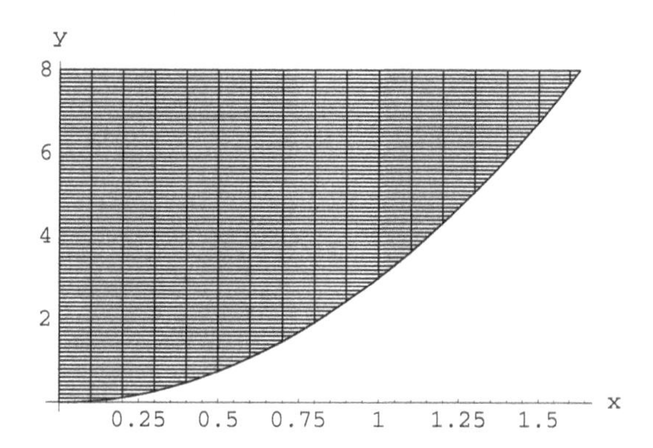

The limits of integration of the second integral describe the region D_2 bounded on the bottom by $y = 8$, on the left by $x = \sqrt{y-8}$ (which is equivalent to $y = x^2 + 8$), and on the right by $x = \sqrt{-y/3}$.

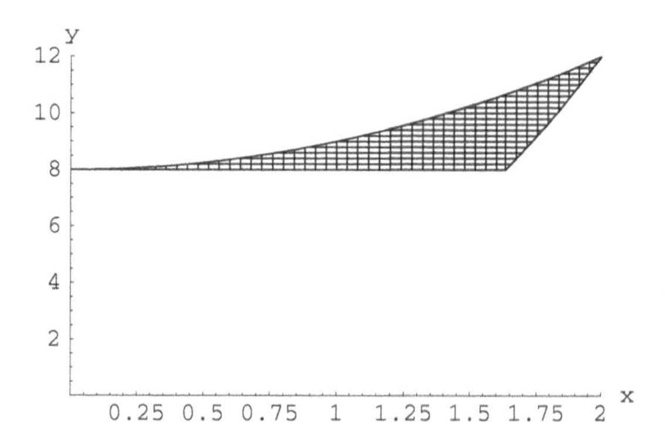

Together, D_1 and D_2 give the full region D of integration.

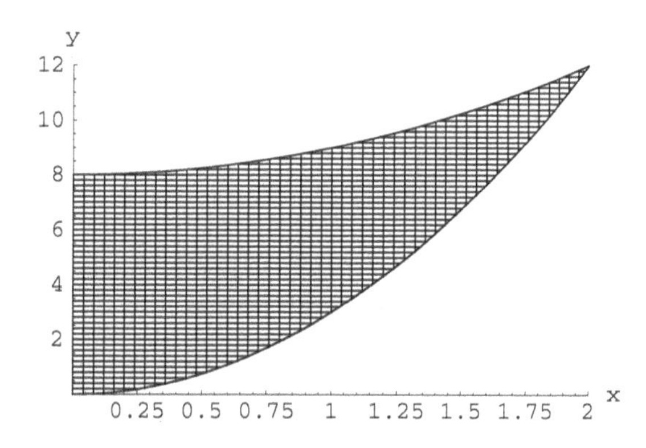

When we reverse the order of integration, the sum of integrals is equal to

$$\begin{aligned}\int_0^2 \int_{3x^2}^{x^2+8} y\,dy\,dx &= \int_0^2 \frac{1}{2}((x^2+8)^2 - 9x^4)\,dx \\ &= \frac{1}{2}\int_0^2 (-8x^4 + 16x^2 + 64)\,dx \\ &= \frac{1}{2}\left(-\frac{256}{5} + \frac{128}{3} + 128\right) = \frac{896}{15}.\end{aligned}$$

17. We reverse the order of integration:

$$\int_0^3 \int_0^{9-x^2} \frac{xe^{3y}}{9-y}\, dy\, dx = \int_0^9 \int_0^{\sqrt{9-y}} \frac{xe^{3y}}{9-y}\, dx\, dy = \int_0^9 \left. \frac{x^2 e^{3y}}{2(9-y)} \right|_0^{\sqrt{9-y}} dy$$

$$= \int_0^9 (e^{3y}/2)\, dx = (e^{3y}/6)\Big|_0^9 = \frac{e^{27}-1}{6}.$$

5.4 Triple Integrals

1. If we integrate with respect to x first, the integral simplifies:

$$\iiint_{[-1,1]\times[0,2]\times[1,3]} xyz\, dV = \int_1^3 \int_0^2 \int_{-1}^1 xyz\, dx\, dy\, dz = \int_1^3 \int_0^2 \left. \frac{x^2 yz}{2} \right|_{-1}^1 dy\, dz$$

$$= \int_1^3 \int_0^2 0\, dy\, dz = 0.$$

5.

$$\int_{-1}^2 \int_1^{z^2} \int_0^{y+z} 3yz^2\, dx\, dy\, dz = \int_{-1}^2 \int_1^{z^2} 3xyz^2 \Big|_0^{y+z} dy\, dz = 3\int_{-1}^2 \int_1^{z^2} (y^2 z^2 + yz^3)\, dy\, dz$$

$$= 3\int_{-1}^2 \left(\frac{y^3 z^2}{3} + \frac{y^2 z^3}{2} \right)\Bigg|_1^{z^2} dz = 3\int_{-1}^2 \left(\frac{z^8}{3} + \frac{z^7}{2} - \frac{z^3}{2} - \frac{z^2}{3} \right) dz$$

$$= 3\left(\frac{z^9}{27} + \frac{z^8}{16} - \frac{z^4}{8} - \frac{z^3}{9} \right)\Bigg|_{-1}^2 = \frac{1539}{16}.$$

9. Of course there are other ways to calculate the volume of the sphere.

$$\text{Volume} = \int_{-a}^a \int_{-\sqrt{a^2-x^2}}^{\sqrt{a^2-x^2}} \int_{-\sqrt{a^2-x^2-y^2}}^{\sqrt{a^2-x^2-y^2}} 1\, dz\, dy\, dx = \int_{-a}^a \int_{-\sqrt{a^2-x^2}}^{\sqrt{a^2-x^2}} 2\sqrt{a^2-x^2-y^2}\, dy\, dx$$

$$= \int_{-a}^a \left(y\sqrt{a^2-x^2-y^2} - (a^2-x^2)\arcsin\left[\frac{y}{\sqrt{a^2-x^2}}\right] \right)\Bigg|_{-\sqrt{a^2-x^2}}^{\sqrt{a^2-x^2}} dx$$

$$= \pi \int_{-a}^a (a^2-x^2)\, dx = \pi(a^2 x - x^3/3)\Big|_{-a}^a = \frac{4\pi a^3}{3}.$$

13. Here $\int_{-3}^3 \int_{x^2}^9 \int_0^{9-y} 8\,xyz\, dz\, dy\, dx = 0$, because we are integrating an odd function in x over an interval that is symmetric in x (see Exercises 1 and 4).

17.

$$\int_0^3 \int_0^{3-x} \int_{-\sqrt{3-x^2/3}}^{\sqrt{3-x^2/3}} (x+y)\, dz\, dy\, dx = \int_0^3 \int_0^{3-x} (2(x+y)\sqrt{3-x^2/3})\, dy\, dx$$

$$= \int_0^3 ((9-x^2)\sqrt{3-x^2/3})\, dx$$

$$= \left(\sqrt{3-x^2/3}\left(\frac{45x-2x^3}{8} \right) + \frac{81\sqrt{3}}{8}\arcsin(x/3) \right)\Bigg|_0^3$$

$$= \frac{81\sqrt{3}\pi}{16}.$$

21. The volume is given by

$$\begin{aligned}\iiint_W 1\,dV &= \int_0^2\int_0^{2-x}\int_0^{4-x^2} 1\,dz\,dy\,dx\\ &= \int_0^2\int_0^{2-x}(4-x^2)\,dy\,dx = \int_0^2(4-x^2)(2-x)\,dx\\ &= \int_0^2(x^3-2x^2-4x+8)\,dx = \left(\frac{x^4}{4}-\frac{2x^3}{3}-2x^2+8x\right)\Big|_0^2 = \frac{20}{3}.\end{aligned}$$

25. The region looks like a wedge of cheese:

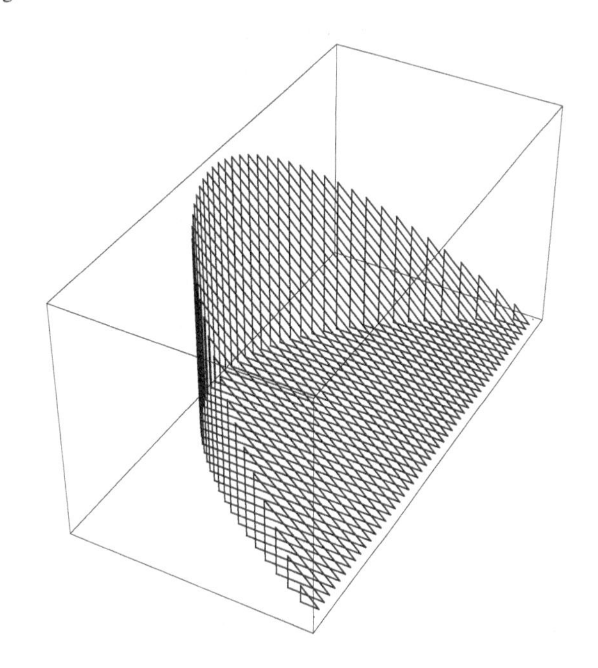

The five other forms are:

$$\begin{aligned}\int_{-1}^1\int_{y^2}^1\int_0^{1-x} f(x,y,z)\,dz\,dx\,dy &= \int_0^1\int_{-\sqrt{x}}^{\sqrt{x}}\int_0^{1-x} f(x,y,z)\,dz\,dy\,dx\\ &= \int_0^1\int_0^{1-x}\int_{-\sqrt{x}}^{\sqrt{x}} f(x,y,z)\,dy\,dz\,dx\\ &= \int_0^1\int_0^{1-z}\int_{-\sqrt{x}}^{\sqrt{x}} f(x,y,z)\,dy\,dx\,dz\\ &= \int_{-1}^1\int_0^{1-y^2}\int_{y^2}^{1-z} f(x,y,z)\,dx\,dz\,dy\\ &= \int_0^1\int_{-\sqrt{1-z}}^{\sqrt{1-z}}\int_{y^2}^{1-z} f(x,y,z)\,dx\,dy\,dz.\end{aligned}$$

29. (a) The solid W is bounded below by the paraboloid $z = x^2+3y^2$, above by the surface $z = 4-y^2$ and in back by the plane $y = 0$. The solid is shown below.

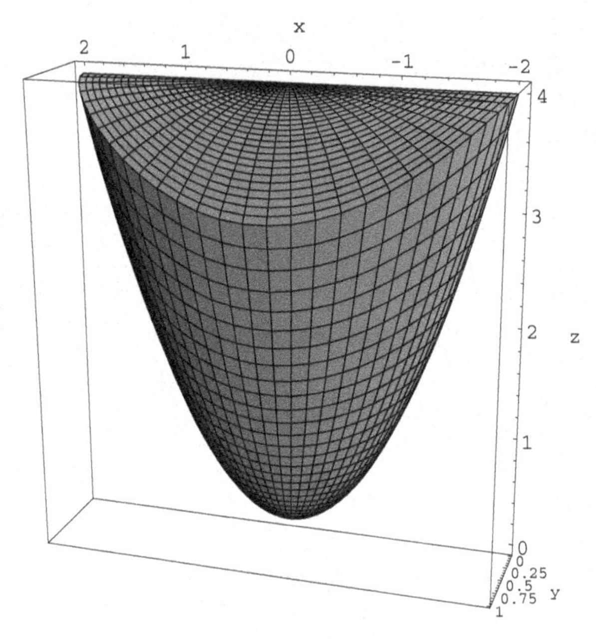

(b) The shadow of W in the xy-plane is half of the region inside the ellipse $x^2 + 4y^2 = 4$ (the half with $y \geq 0$). It may be obtained by finding the intersection curve of $z = x^2 + 3y^2$ and $z = 4 - y^2$ and eliminating z. The shadow looks like

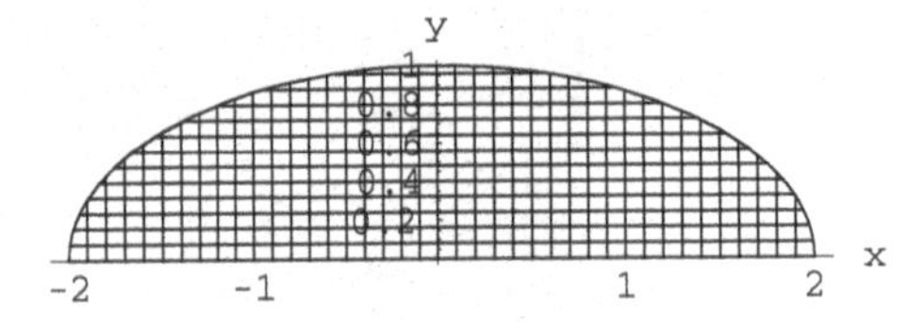

Using the shadow to reverse the order of integration between x and y, we find that the original integral is equivalent to

$$\int_0^1 \int_{-2\sqrt{1-y^2}}^{2\sqrt{1-y^2}} \int_{x^2+3y^2}^{4-y^2} (x^3 + y^3)\, dz\, dx\, dy.$$

(c) We need to consider the shadow of W in the yz-plane.

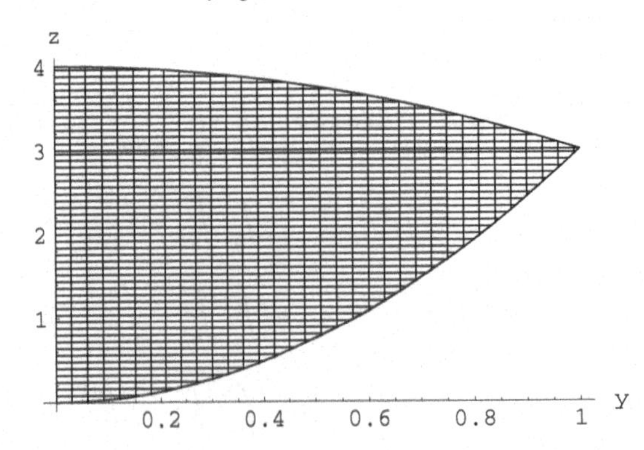

The region is bounded on the left by $y = 0$, on the bottom by $z = 3y^2$ (the section by $x = 0$) and on the top by $z = 4 - y^2$. The full solid W is bounded in the x-direction by the paraboloid $z = x^2 + 3y^2$, which must be expressed in terms of x as $x = \pm\sqrt{z - 3y^2}$. Putting all this information together, we find the desired iterated integral is

$$\int_0^1 \int_{3y^2}^{4-y^2} \int_{-\sqrt{z-3y^2}}^{\sqrt{z-3y^2}} (x^3 + y^3)\, dx\, dz\, dy.$$

(d) Here we use the same shadow in the *yz*-plane as in part (c), only to integrate with respect to y before integrating with respect to z requires dividing the shadow into two regions by the line $z = 3$. (Equivalently, we are dividing the solid W by the plane $z = 3$.) This is why we need a sum of integrals. They are

$$\int_0^3 \int_0^{\sqrt{z/3}} \int_{-\sqrt{z-3y^2}}^{\sqrt{z-3y^2}} (x^3 + y^3)\, dx\, dy\, dz + \int_3^4 \int_0^{\sqrt{4-z}} \int_{-\sqrt{z-3y^2}}^{\sqrt{z-3y^2}} (x^3 + y^3)\, dx\, dy\, dz.$$

(e) To integrate with respect to y first, we need to divide W in a different manner. The shadow in the *xz*-plane shows a region bounded by $z = x^2$ (the section of the paraboloid by $y = 0$) and $z = 4$. However, the curve of intersection of the surfaces $z = x^2 + 3y^2$ and $z = 4 - y^2$ with y eliminated yields the equation $z = \frac{x^2}{4} + 3$. It is along this curve that we must divide the *xz*-shadow and thus the integrals. (Note: This curve is just the shadow of the intersection curve of the two surfaces projected into the *xz*-plane.)

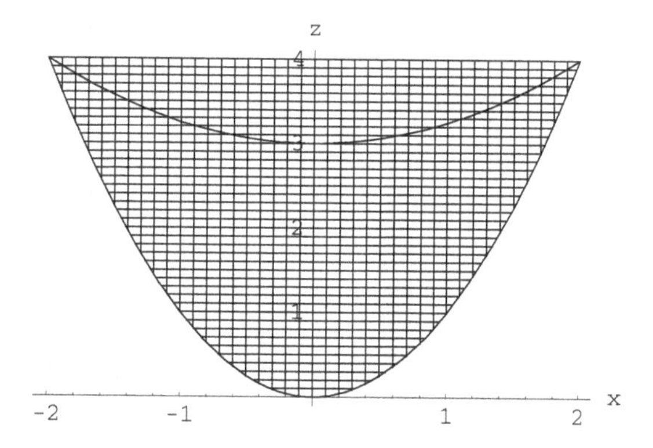

Thus the desired sum of integrals is

$$\int_{-2}^{2} \int_{x^2}^{(x^2/4)+3} \int_0^{\sqrt{(z-x^2)/3}} (x^3 + y^3)\, dy\, dz\, dx + \int_{-2}^{2} \int_{(x^2/4)+3}^{4} \int_0^{\sqrt{4-z}} (x^3 + y^3)\, dy\, dz\, dx.$$

5.5 Change of Variables

1. (a)

$$\mathbf{T}(u, v) = \begin{bmatrix} 3 & 0 \\ 0 & -1 \end{bmatrix} \begin{bmatrix} u \\ v \end{bmatrix}.$$

(b) In this case we can see by inspection that the transformation stretches by 3 in the horizontal direction and reflects without a stretch in the vertical direction. Therefore the image $D = \mathbf{T}(D^*)$ where D^* is the unit square is the rectangle $[0, 3] \times [-1, 0]$.

5. As noted in the text, we have a result for $\mathbf{R}^3$ that is analogous to Proposition 5.1, so as in Exercises 3 and 2 (b) we can just compute the images of the vertices of W^*. We conclude that W^* maps to the parallelepiped with vertices: $(0, 0, 0)$, $(3, 1, 5)$, $(-1, -1, 3)$, $(0, 2, -1)$, $(2, 0, 8)$, $(3, 3, 4)$, $(-1, 1, 2)$, and $(2, 2, 7)$.

11. Here the problem cries out to you to let $u = 2x + y$ and $v = x - y$. Once you've made that move you can easily figure that $\partial(x, y)/\partial(u, v) = -1/3$ and that the new region is $[1, 4] \times [-1, 1]$. So the integral is

$$\int_1^4 \int_{-1}^1 u^2 e^v (1/3)\, dv\, du = \frac{1}{3} \int_1^4 u^2 e^v \Big|_{-1}^{1} du = (e - e^{-1}) \frac{u^3}{9} \Big|_1^4 = 7(e - e^{-1}).$$

15. $\displaystyle \int_0^{2\pi} \int_0^3 r^4\, dr\, d\theta = \int_0^{2\pi} \frac{r^5}{5} \Big|_0^3 d\theta = \int_0^{2\pi} \frac{243}{5}\, d\theta = \frac{486\pi}{5}.$

19. The region in question looks like

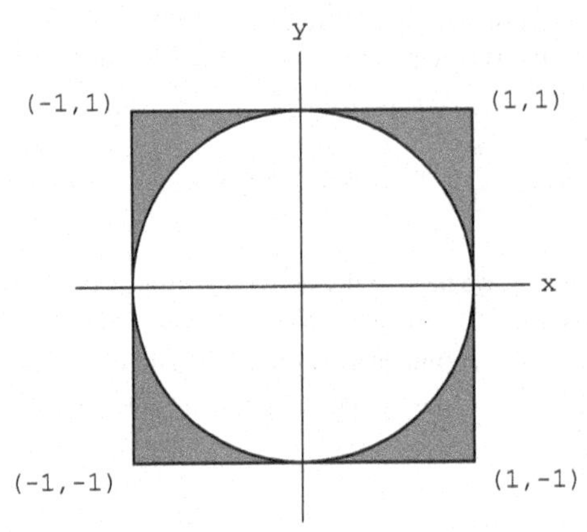

We find $\iint_D y^2 dA = \iint_{\text{square}} y^2 dA - \iint_{\text{disk}} y^2 dA$

$$\iint_{\text{square}} y^2\, dA = \int_{-1}^{1}\int_{-1}^{1} y^2\, dx\, dy = \int_{-1}^{1} 2y^2\, dy = \frac{2}{3}y^3\Big|_{-1}^{1} = \frac{4}{3}$$

$$\iint_{\text{disk}} y^2\, dA = \int_0^{2\pi}\int_0^1 r^2 \sin^2\theta \cdot r\, dr\, d\theta = \int_0^{2\pi} \frac{1}{4}\sin^2\theta\, d\theta$$

$$= \frac{1}{8}\int_0^{2\pi} (1 - \cos 2\theta)\, d\theta = \frac{1}{8}\left(\theta - \frac{1}{2}\sin 2\theta\right)\Big|_0^{2\pi} = \frac{\pi}{4}.$$

Thus

$$\iint_D y^2\, dA = \frac{4}{3} - \frac{\pi}{4} = \frac{16 - 3\pi}{12}.$$

23. We sketch the graphs of the cardioid $r = 1 - \cos\theta$ and the circle $r = 1$:

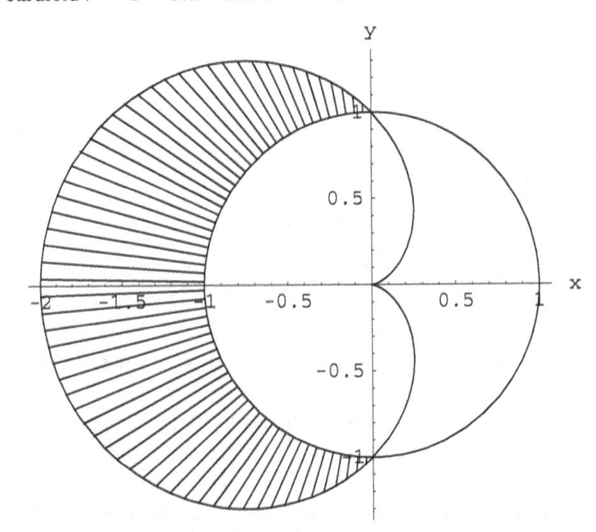

The two curves intersect when $1 - \cos\theta = 1$ which is when $\cos\theta = 0$ so the two points of intersection are $(r, \theta) = (1, \pi/2)$ and $(1, 3\pi/2)$. The region between the two graphs is where $\pi/2 \le \theta \le 3\pi/2$. The area is

$$\int_{\pi/2}^{3\pi/2}\int_1^{1-\cos\theta} r\, dr\, d\theta = \int_{\pi/2}^{3\pi/2}\left(\frac{\cos^2\theta}{2} - \cos\theta\right) d\theta$$

$$= \frac{1}{8}(2\theta - 8\sin\theta + \sin 2\theta)\Big|_{\pi/2}^{3\pi/2} = 2 + \frac{\pi}{4}.$$

27. Two of the edges of the unit square are given by $x = 1$ (or $r = 1/\cos\theta$ in polar coordinates) and by $y = 1$ (i.e., by $r = 1/\sin\theta$). We need to divide the square along the $\theta = \pi/4$ line, and use a sum of integrals:

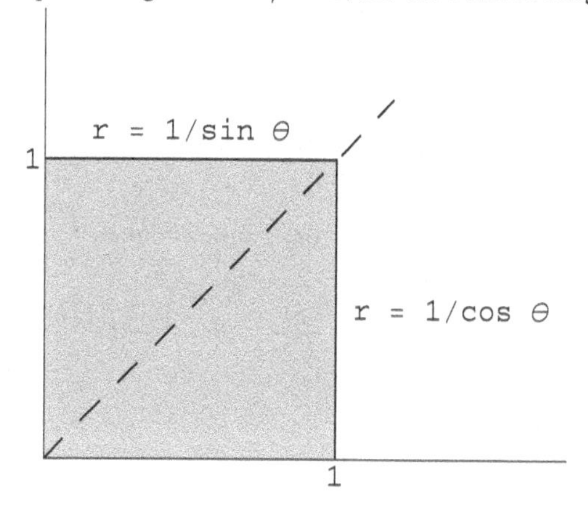

Thus

$$\iint_D \frac{x}{\sqrt{x^2+y^2}}\,dA = \int_0^{\pi/4}\int_0^{1/\cos\theta} \frac{r\cos\theta}{r}\cdot r\,dr\,d\theta + \int_{\pi/4}^{\pi/2}\int_0^{1/\sin\theta} r\cos\theta\,dr\,d\theta$$

$$= \int_0^{\pi/4} \frac{1}{2\cos^2\theta}\cdot\cos\theta\,d\theta + \int_{\pi/4}^{\pi/2} \frac{1}{2\sin^2\theta}\cdot\cos\theta\,d\theta$$

$$= \int_0^{\pi/4} \frac{1}{2}\sec\theta\,d\theta + \int_{\theta=\pi/4}^{\theta=\pi/2} \frac{1}{2(\sin\theta)^2}\,d(\sin\theta)$$

$$= \frac{1}{2}\ln|\sec\theta+\tan\theta|\Big|_0^{\pi/4} - \frac{1}{2\sin\theta}\Big|_{\pi/4}^{\pi/2}$$

$$= \frac{1}{2}\ln(\sqrt{2}+1) - \frac{1}{2}\ln 1 - \frac{1}{2} + \frac{\sqrt{2}}{2} = \frac{1}{2}(\ln(\sqrt{2}+1)+\sqrt{2}-1).$$

31. Here we will use cylindrical coordinates:

$$\iiint_W (x^2+y^2+2z^2)\,dV = \int_{-1}^{2}\int_0^{2\pi}\int_0^{2} r(r^2+2z^2)\,dr\,d\theta\,dz$$

$$= \int_{-1}^{2}\int_0^{2\pi} (4z^2+4)\,d\theta\,dz$$

$$= \int_{-1}^{2} (8\pi z^2+8\pi)\,dz = 48\pi.$$

35. Here we use spherical coordinates.

$$\iiint_W \sqrt{x^2+y^2+z^2}\,e^{x^2+y^2+z^2}\,dV = \int_0^{2\pi}\int_0^{\pi}\int_a^{b} (\rho^3 e^{\rho^2}\sin\varphi)\,d\rho\,d\varphi\,d\theta$$

$$= \frac{1}{2}\int_0^{2\pi}\int_0^{\pi} ([(1-a^2)e^{a^2}+(b^2-1)e^{b^2}]\sin\varphi)\,d\varphi\,d\theta$$

$$= \int_0^{2\pi} ((1-a^2)e^{a^2}+(b^2-1)e^{b^2})\,d\theta$$

$$= 2\pi((1-a^2)e^{a^2}+(b^2-1)e^{b^2}).$$

37. We use spherical coordinates, in which case the cone $z = \sqrt{3x^2+3y^2}$ has equation

$$\rho\cos\varphi = \sqrt{3}\rho\sin\varphi \iff \tan\varphi = \frac{1}{\sqrt{3}} \iff \varphi = \frac{\pi}{6}.$$

The sphere $x^2+y^2+z^2=6z$ has spherical equation

$$\rho^2 = 6\rho\cos\varphi \iff \rho = 6\cos\varphi.$$

Thus

$$\begin{aligned}\iiint_W z^2\,dV &= \int_0^{2\pi}\int_0^{\pi/6}\int_0^{6\cos\varphi}(\rho^2\cos^2\varphi)\cdot\rho^2\sin\varphi\,d\rho\,d\varphi\,d\theta \\ &= \int_0^{2\pi}\int_0^{\pi/6}\frac{6^5}{5}\cos^7\varphi\sin\varphi\,d\varphi\,d\theta = \int_0^{2\pi}\left(-\frac{7776}{40}\cos^8\varphi\right)\Big|_{\varphi=0}^{\varphi=\pi/6}\,d\theta \\ &= \int_0^{2\pi}\frac{972}{5}\left(1-\frac{81}{256}\right)d\theta = \frac{972}{5}\left(\frac{175}{128}\right)\pi = \frac{8505\pi}{32}.\end{aligned}$$

41. We will again use cylindrical coordinates.

$$\begin{aligned}\iiint_W (2+x^2+y^2)\,dV &= \int_3^5\int_0^{2\pi}\int_0^{\sqrt{25-z^2}}(2+r^2)r\,dr\,d\theta\,dz \\ &= \int_3^5\int_0^{2\pi}\left(\frac{z^4}{4}-\frac{27z^2}{2}+\frac{725}{4}\right)d\theta\,dz \\ &= \int_3^5\left(2\pi\left[\frac{z^4}{4}-\frac{27z^2}{2}+\frac{725}{4}\right]\right)dz = \frac{656\pi}{5}.\end{aligned}$$

5.6 Applications of Integration

1. (a) Let's assume a 30-day month.

$$\begin{aligned}[f]_{\text{avg}} &= \frac{1}{30}\int_0^{30} I(x)\,dx = \frac{1}{30}\int_0^{30}\left(75\cos\frac{\pi x}{15}+80\right)dx \\ &= \frac{1}{30}\left(\frac{1125}{\pi}\sin\frac{\pi x}{15}+80x\right)\Big|_0^{30} = 2400/30 = 80 \text{ cases.}\end{aligned}$$

(b) Here the 2 cents will be a constant that pulls through the integral so the average holding cost is just 2 cents times the average daily inventory, or \$1.60.

5. (a) We are told that in the $2\times 2\times 2$ cube centered at the origin, $T(x,y,z)=c(x^2+y^2+z^2)$. The average temperature of the cube is

$$\begin{aligned}[T]_{\text{avg}} &= \frac{c}{8}\int_{-1}^{1}\int_{-1}^{1}\int_{-1}^{1}(x^2+y^2+z^2)\,dx\,dy\,dz = \frac{c}{8}\int_{-1}^{1}\int_{-1}^{1}(2z^2+2y^2+2/3)\,dy\,dz \\ &= \frac{c}{8}\int_{-1}^{1}(4z^2+8/3)\,dz = \frac{c}{8}(8) = c.\end{aligned}$$

(b) $T(x,y,z)=c$ when $x^2+y^2+z^2=1$ so the temperature is equal to the average temperature on the surface of the unit sphere.

9. This is an extension of Exercise 8. The domain is $[0,6]\times[0,6]\times[0,6]$. This time there is six-fold symmetry so we will calculate the average value for z in the region where $z\le y\le x$ and multiply by 6 and then divide by 6^3 which is the volume of the domain.

$$\begin{aligned}[\text{Time}]_{\text{avg}} &= \frac{6}{216}\int_0^6\int_0^x\int_0^y z\,dz\,dy\,dx = \frac{1}{36}\int_0^6\int_0^x\frac{y^2}{2}\,dy\,dz \\ &= \frac{1}{36}\int_0^6\frac{x^3}{6}\,dx = \frac{1}{36}\left(\frac{x^4}{24}\right)\Big|_0^6 = 3/2.\end{aligned}$$

So with three train lines the average wait is 90 seconds.

13. We first calculate

$$M = \int_0^9 \int_0^{\sqrt{x}} (xy)\, dy\, dx = \frac{243}{2}$$

$$M_y = \int_0^9 \int_0^{\sqrt{x}} (x^2 y)\, dy\, dx = \frac{6561}{8}$$

$$M_x = \int_0^9 \int_0^{\sqrt{x}} (xy^2)\, dy\, dx = \frac{1458}{7}$$

so

$$\bar{x} = \frac{6561/8}{243/2} = 27/4 \quad \text{and} \quad \bar{y} = \frac{1458/7}{243/2} = 12/7.$$

17. The region in question looks as follows:

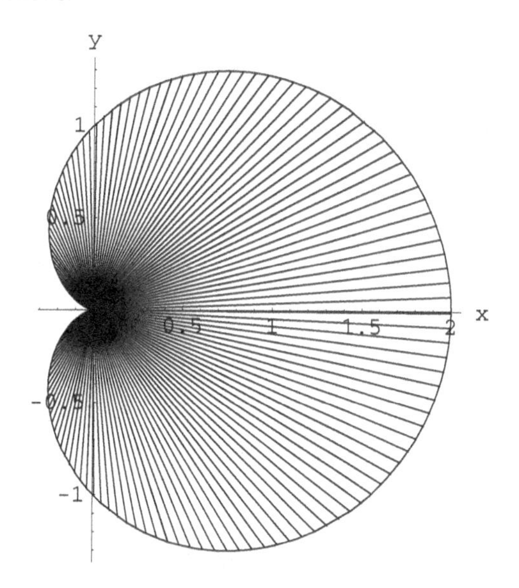

$$\text{Total mass } M = \iint_D \delta\, dA = \int_0^{2\pi} \int_0^{1+\cos\theta} r^2 dr\, d\theta$$

$$= \int_0^{2\pi} \frac{1}{3}(1+\cos\theta)^3 d\theta$$

$$= \int_0^{2\pi} \frac{1}{3}(1 + 3\cos\theta + 3\cos^2\theta + \cos^3\theta)\, d\theta = \frac{5\pi}{3}$$

$$M_y = \iint_D x\delta\, dA = \int_0^{2\pi} \int_0^{1+\cos\theta} r^3 \cos\theta\, dr\, d\theta = \int_0^{2\pi} \frac{1}{4}(1+\cos\theta)^4 \cos\theta\, d\theta$$

$$= \frac{7\pi}{4} \qquad \text{(after some effort!)}$$

$$M_x = \iint_D y\delta\, dA = \int_0^{2\pi} \int_0^{1+\cos\theta} r^3 \sin\theta\, dr\, d\theta$$

$$= \int_0^{2\pi} \frac{1}{4}(1+\cos\theta)^4 \sin\theta\, d\theta = -\frac{1}{4}\int_2^2 u^4 du = 0$$

(It's also possible to see this from symmetry.) Thus

$$\bar{x} = \frac{7\pi}{4} \cdot \frac{3}{5\pi} = \frac{21}{20}, \qquad \bar{y} = 0.$$

19. **(a)** First calculate:

$$M = \int_{-1}^{2}\int_{-1}^{1}\int_{3y^2}^{3} dz\,dy\,dx = 12$$

$$M_{yz} = \int_{-1}^{2}\int_{-1}^{1}\int_{3y^2}^{3} x\,dz\,dy\,dx = 6$$

$$M_{xz} = \int_{-1}^{2}\int_{-1}^{1}\int_{3y^2}^{3} y\,dz\,dy\,dx = 0$$

$$M_{xy} = \int_{-1}^{2}\int_{-1}^{1}\int_{3y^2}^{3} z\,dz\,dy\,dx = \frac{108}{5}.$$

This means that $(\bar{x}, \bar{y}, \bar{z}) = (1/2, 0, 9/5)$.

(b) Next we calculate the center of mass with the given density function.

$$M = \int_{-1}^{2}\int_{-1}^{1}\int_{3y^2}^{3} (z + x^2)\,dz\,dy\,dx = \frac{168}{5}$$

$$M_{yz} = \int_{-1}^{2}\int_{-1}^{1}\int_{3y^2}^{3} x(z + x^2)\,dz\,dy\,dx = \frac{129}{5}$$

$$M_{xz} = \int_{-1}^{2}\int_{-1}^{1}\int_{3y^2}^{3} y(z + x^2)\,dz\,dy\,dx = 0$$

$$M_{xy} = \int_{-1}^{2}\int_{-1}^{1}\int_{3y^2}^{3} z(z + x^2)\,dz\,dy\,dx = \frac{2376}{35}.$$

This means that $(\bar{x}, \bar{y}, \bar{z}) = (43/56, 0, 99/49)$.

23. By symmetry $\bar{x} = \bar{y} = \bar{z}$. $\bar{z}$ is easiest to find. Volume of W is $\frac{1}{8}\left(\frac{4}{3}\pi a^3\right) = \frac{\pi a^3}{6}$.

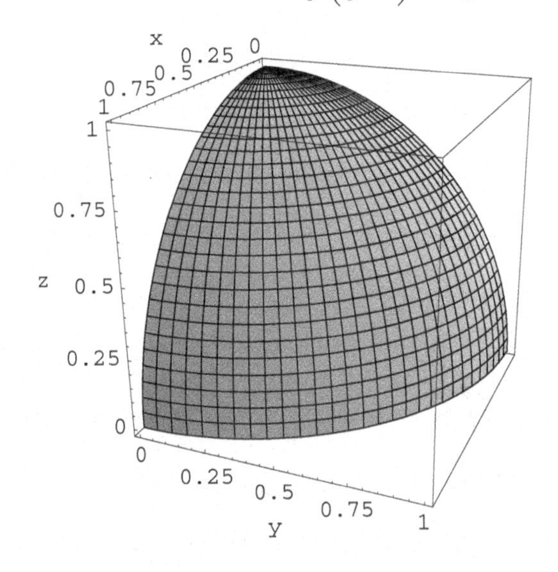

Thus

$$\bar{z} = \frac{6}{\pi a^3}\iiint_W z\,dV = \frac{6}{\pi a^3}\int_0^{\pi/2}\int_0^{\pi/2}\int_0^a \rho\cos\varphi\cdot\rho^2\sin\varphi\,d\rho\,d\varphi\,d\theta$$

$$= \frac{6}{\pi a^3}\int_0^{\pi/2}\int_0^{\pi/2}\frac{a^4}{4}\cos\varphi\sin\varphi\,d\varphi\,d\theta = \frac{3a}{2\pi}\int_0^{\pi/2}\frac{1}{2}d\theta$$

$$= \frac{3a}{8}.$$

25. **(a)** By symmetry we can see that the moment of inertia about each of the coordinate axes is the same.

$$I_x = I_y = I_z = \int_0^1 \int_0^{1-x} \int_0^{1-x-y} (y^2 + z^2)\, dz\, dy\, dx = \frac{1}{30}.$$

(b) Again, by symmetry we see that the radius of gyration about each of the coordinate axes is the same. We calculate

$$M = \int_0^1 \int_0^{1-x} \int_0^{1-x-y} dz\, dy\, dx = \frac{1}{6}.$$

Then $r_x = r_y = r_z = \sqrt{\dfrac{1/30}{1/6}} = 1/\sqrt{5}$.

27. **(a)** The problem cries out to be solved using cylindrical coordinates. For I_z this means that the $x^2 + y^2$ in the integrand is r^2 so

$$\begin{aligned} I_z &= \int_0^3 \int_0^{2\pi} \int_{r^2}^9 2zr^3\, dz\, d\theta\, dr = \frac{6561\pi}{4} \quad \text{and} \\ M &= \int_0^3 \int_0^{2\pi} \int_{r^2}^9 2zr\, dz\, d\theta\, dr = 486\pi, \quad \text{so} \\ r_z &= \frac{3\sqrt{3}}{2\sqrt{2}}. \end{aligned}$$

(b) This time

$$\begin{aligned} I_z &= \int_0^3 \int_0^{2\pi} \int_{r^2}^9 r^4\, dz\, d\theta\, dr = \frac{8748\pi}{35} \quad \text{and} \\ M &= \int_0^3 \int_0^{2\pi} \int_{r^2}^9 r^2\, dz\, d\theta\, dr = \frac{324\pi}{5}, \quad \text{so} \\ r_z &= \frac{3\sqrt{3}}{\sqrt{7}}. \end{aligned}$$

31. The only adjustment in the formula for I_x is because we are using the square of the distance from the line $y = 3$ and not the formula given in text which squares the distance from the x-axis. This is a straightforward application of formula (8).

$$I_{y=3} = \int_{-2}^2 \int_{-\sqrt{4-x^2}}^{\sqrt{4-x^2}} (x^2(3-y)^2))\, dy\, dx = \frac{116\pi}{3}.$$

5.7 Numerical Approximations of Multiple Integrals

1. **(a)** We let $\Delta x = \dfrac{3.1 - 3}{2} = 0.05$, $\Delta y = \dfrac{2.1 - 1.5}{3} = 0.2$. Thus, using formula (6) with $f(x, y) = x^2 - 6y^2$, we have

$$\begin{aligned} T_{2,3} = \frac{(0.05)(0.2)}{4} & [f(3, 1.5) + f(3, 2.1) + f(3.1, 1.5) + f(3.1, 2.1) \\ & + 2\,(f(3, 1.7) + f(3, 1.9) + f(3.05, 1.5) + f(3.05, 2.1) + f(3.1, 1.7) + f(3.1, 1.9)) \\ & + 4\,(f(3.05, 1.7) + f(3.05, 1.9))] \\ = & -0.621375 \end{aligned}$$

(b) $\displaystyle\int_3^{3.1} \int_{1.5}^{2.1} (x^2 - 6y^2)\, dy\, dx = \int_3^{3.1} (x^2 y - 2y^3)\Big|_{y=1.5}^{y=2.1} dx = \int_3^{3.1} \left[0.6x^2 - 2(2.1^3 - 1.5^3)\right] dx$

$\displaystyle = \left[0.2x^3 - 2(2.1^3 - 1.5^3)x\right]\Big|_3^{3.1} = 0.2(3.1^3 - 3^3) - 0.2(2.1^3 - 1.5^3) = -0.619$

5. (a) We let $\Delta x = \dfrac{1.1-1}{2} = 0.05$, $\Delta y = \dfrac{0.6-0}{3} = 0.2$. Thus, using formula (6) with $f(x,y) = e^{x+2y}$, we have

$$\begin{aligned} T_{2,3} = {} & \frac{(0.05)(0.2)}{4}\left[f(1,0) + f(1,0.6) + f(1.1,0) + f(1.1,0.6)\right. \\ & + 2\left(f(1,0.2) + f(1,0.4) + f(1.05,0) + f(1.05,0.6) + f(1.1,0.2) + f(1.1,0.4)\right) \\ & \left. + 4\left(f(1.05,0.2) + f(1.05,0.4)\right)\right] \\ = {} & 0.336123 \end{aligned}$$

(b) $\displaystyle\int_1^{1.1}\int_0^{0.6} e^{x+2y}\,dy\,dx = \int_1^{1.1}\int_0^{0.6} e^x e^{2y}\,dy\,dx = \int_1^{1.1} e^x\left(\frac{1}{2}e^{2y}\right)\Bigg|_{y=0}^{y=0.6} dx$

$\displaystyle = \int_1^{1.1}\left(\frac{e^{1.2}-1}{2}\right)e^x\,dx = \frac{e^{1.2}-1}{2}(e^{1.1}-1) = 0.331642$

9. (a) We let $\Delta x = \dfrac{1.1-1}{2} = 0.05$, $\Delta y = \dfrac{0.6-0}{2} = 0.3$. Hence, with $f(x,y) = e^{x+2y}$, we have

$$\begin{aligned} S_{2,2} = {} & \frac{(0.05)(0.3)}{9}\left[f(1,0) + f(1,0.6) + f(1.1,0) + f(1.1,0.6)\right. \\ & \left. + 4\left(f(1,0.3) + f(1.05,0) + f(1.05,0.6) + f(1.1,0.3)\right) + 16f(1.05,0.3)\right] \\ = {} & 0.331871 \end{aligned}$$

(b) In part (b) of Exercise 5 we calculated $\displaystyle\int_1^{1.1}\int_0^{0.6} e^{x+2y}\,dy\,dx$ to be 0.331642.

13. (a) The paraboloid is a portion of the graph of $f(x,y) = 4 - x^2 - 3y^2$. We have $\partial f/\partial x = -2x$, $\partial f/\partial y = -6y$ so that the surface area integral we desire is

$$\int_0^1\int_0^1 \sqrt{4x^2+36y^2+1}\,dy\,dx.$$

(b) We let $\Delta x = \Delta y = \dfrac{1-0}{4} = 0.25$. With $g(x,y) = \sqrt{4x^2+36y^2+1}$, we have

$$\begin{aligned} T_{4,4} = {} & \frac{(0.25)(0.25)}{4}\left[g(0,0) + g(0,1) + g(1,0) + g(1,1)\right. \\ & + 2\left(g(0,0.25) + g(0,0.5) + g(0,0.75) + g(0.25,0) + g(0.25,1) + g(0.5,0)\right. \\ & \quad \left. + g(0.5,1) + g(0.75,0) + g(0.75,1) + g(1,0.25) + g(1,0.5) + g(1,0.75)\right) \\ & + 4\left(g(0.25,0.25) + g(0.25,0.5) + g(0.25,0.75) + g(0.5,0.25) + g(0.5,0.5)\right. \\ & \quad \left.\left. + g(0.5,0.75) + g(0.75,0.25) + g(0.75,0.5) + g(0.75,0.75)\right)\right] \\ = {} & 3.52366 \end{aligned}$$

17. (a) We have

$$\frac{\partial^2}{\partial x^2}\left(e^{x-y}\right) = \frac{\partial^2}{\partial y^2}\left(e^{x-y}\right) = e^{x-y}.$$

The maximum value of e^{x-y} on $[0,0.3]\times[0,0.4]$ is $e^{0.3-0} = e^{0.3}$. Furthermore, in computing the approximation $T_{n,n}$ we have $\Delta x = 0.3/n$ and $\Delta y = 0.4/n$. Thus Theorem 7.3 implies that

$$|E_{n,n}| \le \frac{(0.3)(0.4)}{12}\left[\left(\frac{0.3}{n}\right)^2 e^{0.3} + \left(\frac{0.4}{n}\right)^2 e^{0.3}\right] = \frac{(0.3)(0.4)(0.5)^2 e^{0.3}}{12n^2}.$$

For this expression to be at most 10^{-5}, we must have

$$\frac{(0.3)(0.4)(0.5)^2 e^{0.3}}{12n^2} \le 10^{-5} \iff n^2 \ge \frac{10^5(0.3)(0.4)(0.5)^2 e^{0.3}}{12} \iff n > 18.37.$$

Thus we should take n to be at least 19.

(b) In this case, we use Theorem 7.4. First note that we have

$$\frac{\partial^4}{\partial x^4}\left(e^{x-y}\right) = \frac{\partial^4}{\partial y^4}\left(e^{x-y}\right) = e^{x-y},$$

so that, as in part (a), the maximum value of e^{x-y} on $[0, 0.3] \times [0, 0.4]$ is $e^{0.3}$. Moreover, in computing the approximation $S_{2n,2n}$, we have $\Delta x = 0.3/(2n)$ and $\Delta y = 0.4/(2n)$. Therefore, Theorem 7.4 implies that

$$|E_{2n,2n}| \le \frac{(0.3)(0.4)}{180}\left[\left(\frac{0.3}{2n}\right)^4 e^{0.3} + \left(\frac{0.4}{2n}\right)^4 e^{0.3}\right] = \frac{(0.3)(0.4)((0.3)^4 + (0.4)^4)e^{0.3}}{180 \cdot 16n^4}.$$

For this expression to be at most 10^{-5}, we must have

$$\frac{(0.3)(0.4)((0.3)^4 + (0.4)^4)e^{0.3}}{180 \cdot 16n^4} \le 10^{-5} \iff n^4 \ge \frac{10^5(0.3)(0.4)((0.3)^4 + (0.4)^4)e^{0.3}}{180 \cdot 16}$$
$$\iff n > 0.659.$$

Thus, since n must be an integer, we must have n at least 1; that is, $S_{2,2}$ will give an approximation with the desired accuracy.

21. We let $\Delta x = \dfrac{\pi/4 - 0}{3} = \dfrac{\pi}{12}$, so that $x_0 = 0$, $x_1 = \frac{\pi}{12}$, $x_2 = \frac{\pi}{6}$, $x_3 = \frac{\pi}{4}$. Then $\Delta y(x) = \dfrac{\cos x - \sin x}{3}$, so that

$$\Delta y(0) = \tfrac{1}{3} \implies y_0(x_0) = 0,\ y_1(x_0) = \tfrac{1}{3},\ y_2(x_0) = \tfrac{2}{3},\ y_3(x_0) = 1$$

$$\Delta y\left(\tfrac{\pi}{12}\right) = \tfrac{1}{3\sqrt{2}} \quad \text{(by use of the half-angle formula)}$$

$$\implies y_0(x_1) = \sin\left(\tfrac{\pi}{12}\right),\ y_1(x_1) = \sin\left(\tfrac{\pi}{12}\right) + \tfrac{1}{3\sqrt{2}},$$
$$y_2(x_1) = \sin\left(\tfrac{\pi}{12}\right) + \tfrac{2}{3\sqrt{2}},\ y_3(x_1) = \sin\left(\tfrac{\pi}{12}\right) + \tfrac{1}{\sqrt{2}}$$

$$\Delta y\left(\tfrac{\pi}{6}\right) = \tfrac{\sqrt{3}-1}{6} \implies y_0(x_2) = \tfrac{1}{2},\ y_1(x_2) = \tfrac{\sqrt{3}+2}{6},\ y_2(x_2) = \tfrac{2\sqrt{3}+1}{6},\ y_3(x_2) = \tfrac{\sqrt{3}}{2}$$

$$\Delta y\left(\tfrac{\pi}{4}\right) = 0 \implies \text{partition points not needed.}$$

This information is pictured in the figure below.

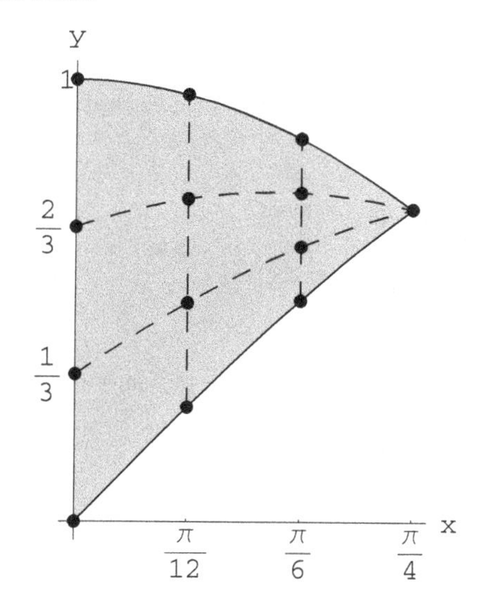

Thus, using $f(x, y) = 2x \cos y + \sin^2 x$, we have

$$T_{3,3} = \frac{(\pi/12)(1/3)}{4}\left[f(0,0) + 2f(0,\tfrac{1}{3}) + 2f(0,\tfrac{2}{3}) + f(0,1)\right]$$

$$+ \frac{(\pi/12)(1/(3\sqrt{2}))}{4}\Big[2f(\tfrac{\pi}{12}, \sin\tfrac{\pi}{12}) + 4f(\tfrac{\pi}{12}, \sin\tfrac{\pi}{12} + \tfrac{1}{3\sqrt{2}})$$

$$+ 4f(\tfrac{\pi}{12}, \sin\tfrac{\pi}{12} + \tfrac{2}{3\sqrt{2}}) + 2f(\tfrac{\pi}{12}, \cos\tfrac{\pi}{12})\Big]$$

$$+ \frac{(\pi/12)((\sqrt{3}-1)/6)}{4}\Big[2f(\tfrac{\pi}{6}, \tfrac{1}{2}) + 4f(\tfrac{\pi}{6}, \tfrac{\sqrt{3}+2}{6}) + 4f(\tfrac{\pi}{6}, \tfrac{2\sqrt{3}+1}{6}) + 2f(\tfrac{\pi}{6}, \tfrac{\sqrt{3}}{2})\Big]$$

$$= 0.190978$$

(This approximation turns out to be rather low.)

25. We let $\Delta y = \dfrac{1.6 - 1}{3} = 0.2$, so that $y_0 = 1$, $y_1 = 1.2$, $y_2 = 1.4$, $y_3 = 1.6$. Then $\Delta x(y) = \dfrac{2y - y}{3} = \dfrac{y}{3}$, so that

$$\Delta x(1) = \tfrac{1}{3} \implies x_0(y_0) = 1,\ x_1(y_0) = \tfrac{4}{3},\ x_2(y_0) = \tfrac{5}{3},\ x_3(y_0) = 2$$

$$\Delta x(1.2) = 0.4 \implies x_0(y_1) = 1.2,\ x_1(y_1) = 1.6,\ x_2(y_1) = 2,\ x_3(y_1) = 2.4$$

$$\Delta x(1.4) = 0.4\overline{6} \implies x_0(y_2) = 1.4,\ x_1(y_2) = 1.8\overline{6},\ x_2(y_2) = 2.\overline{3},\ x_3(y_2) = 2.8$$

$$\Delta x(1.6) = 0.5\overline{3} \implies x_0(y_3) = 1.6,\ x_1(y_3) = 2.1\overline{3},\ x_2(y_3) = 2.\overline{6},\ x_3(y_3) = 3.2$$

Thus, using $f(x, y) = \ln(xy)$, we have

$$T_{3,3} = \frac{(0.2)(0.\overline{3})}{4}\left[f(1,1) + 2f(\tfrac{4}{3},1) + 2f(\tfrac{5}{3},1) + f(2,1)\right]$$

$$+ \frac{(0.2)(0.4)}{4}\left[2f(1.2,1.2) + 4f(1.6,1.2) + 4f(2,1.2) + 2f(2.4,1.2)\right]$$

$$+ \frac{(0.2)(0.4\overline{6})}{4}\left[2f(1.4,1.4) + 4f(1.8\overline{6},1.4) + 4f(2.\overline{3},1.4) + 2f(2.8,1.4)\right]$$

$$+ \frac{(0.2)(0.5\overline{3})}{4}\left[f(1.6,1.6) + 2f(2.1\overline{3},1.6) + 2f(2.\overline{6},1.6) + f(3.2,1.6)\right]$$

$$= 0.724061$$

(This actual value is closer to 0.724519.)

True/False Exercises for Chapter 5

1. False. (Not all rectangles must have sides parallel to the coordinate axes.)
3. True.
5. False. (Let $f(x, y) = x$, for example.)
7. False. (The integral on the right isn't even a number!)
9. True.
11. True.
13. False. (The value of the integral is 3.)
15. True.
17. True. (The inner integral with respect to z is zero because of symmetry.)
19. True.
21. False. (The integrals are opposites of one another.)
23. False. (A factor of r should appear in the integrand.)
25. False. (A factor of ρ is missing in the integrand.)
27. True.
29. True.

Miscellaneous Exercises for Chapter 5

1. First let's split the integrand:

$$\iiint_B (z^3 + 2)\, dV = \iiint_B z^3 dV + \iiint_B 2\, dV = \iiint_B z^3 dV + 2\iiint_B dV.$$

Here B is the ball of radius 3 centered at the origin. The integral of an odd function of z over a region which is symmetric with respect to z is 0. The other integral is twice the volume of a sphere of radius 3 so

$$\iiint_B (z^3 + 2)\, dV = 2\left(\frac{4}{3}\pi 3^3\right) = 72\pi.$$

5. I think the given form is the easiest to integrate:

$$\begin{aligned}\int_0^{2\pi}\int_0^1\int_0^{\sqrt{9-r^2}} r\, dz\, dr\, d\theta &= \int_0^{2\pi}\int_0^1 r\sqrt{9-r^2}\, dr\, d\theta \\ &= \int_0^{2\pi}\left(9 - \frac{16\sqrt{2}}{3}\right) d\theta \\ &= 2\pi\left(9 - \frac{16\sqrt{2}}{3}\right).\end{aligned}$$

(a) In Cartesian coordinates, z doesn't really change and for the outer two limits, we are integrating over a unit circle so our answer is

$$\int_{-1}^1\int_{-\sqrt{1-x^2}}^{\sqrt{1-x^2}}\int_0^{\sqrt{9-x^2-y^2}} dz\, dy\, dx.$$

(b) The solid is the intersection of the top half of a sphere of radius 3 centered at the origin and a cylinder of radius 1 with axis of symmetry the z-axis. In spherical coordinates this means that we have to split the integral into two pieces: one that corresponds to the spherical cap and one that corresponds to the straight sides. The "cone" of intersection is when $\varphi = \sin^{-1} 1/3$. For the integral that corresponds to the "straight sides", $0 \leq r \leq 1$. In spherical coordinates that is $0 \leq \rho\sin\varphi \leq 1$ or $0 \leq \rho \leq \csc\varphi$. The integrals are, therefore,

$$\int_0^{2\pi}\int_0^{\sin^{-1} 1/3}\int_0^3 \rho^2\sin\varphi\, d\rho\, d\varphi\, d\theta + \int_0^{2\pi}\int_{\sin^{-1} 1/3}^{\pi/2}\int_0^{\csc\varphi} \rho^2\sin\varphi\, d\rho\, d\varphi\, d\theta.$$

7. Orient the cube so that a vertex is at the origin and the edges that meet at that vertex lie along the x-, y- and z-axes so that the cube is in the first octant. We'll double the volume of half of the cube. In this case $0 \leq z \leq a$, $0 \leq \theta \leq \pi/4$ and the only difficulty is with r. The radius varies from 0 to the line $x = a$. In cylindrical coordinates $x = r\cos\theta$ so $r = a\sec\theta$ and our limits for r are $0 \leq r \leq a\sec\theta$. The volume is

$$V = 2\int_0^a\int_0^{\pi/4}\int_0^{a\sec\theta} r\, dr\, d\theta\, dz = \int_0^a\int_0^{\pi/4} a^2\sec^2\theta\, d\theta\, dz = \int_0^a a^2\, dz = a^3.$$

The above calculation wouldn't change much if you followed the hint in the text and placed the center of the cube at the origin. In this case you would have 1/8 of the figure in the first octant and you would be calculating the volume of a cube with sides $a/2$.

11. (a) As we've seen before, we can write the integral as $\displaystyle\int_{-a}^a\int_{-b\sqrt{1-x^2/a^2}}^{b\sqrt{1-x^2/a^2}} dy\, dx.$

(b) When we scale by letting $x = a\bar{x}$ and $y = b\bar{y}$, the ellipse is transformed into the unit circle E^* in the $\bar{x}\bar{y}$-plane. To rewrite the integral we also quickly calculate that $\partial(x,y)/\partial(\bar{x},\bar{y}) = ab$. The transformed integral is $\displaystyle\int_{-1}^1\int_{-\sqrt{1-\bar{x}^2}}^{\sqrt{1-\bar{x}^2}} ab\, d\bar{y}\, d\bar{x}.$

(c) Because we are integrating over a unit circle, we transform to polar coordinates (o.k., really we do it because the text tells us to):

$$\int_{-1}^1\int_{-\sqrt{1-\bar{x}^2}}^{\sqrt{1-\bar{x}^2}} ab\, d\bar{y}\, d\bar{x} = \int_0^{2\pi}\int_0^1 abr\, dr\, d\theta = \int_0^{2\pi}\frac{1}{2}ab\, d\theta = \frac{1}{2}ab(2\pi) = \pi\, ab.$$

15. If you didn't first sketch the region you may be tempted to use the numerator and denominator of the integrand as your new variables. The diamond-like shape is bounded on two sides by the hyperbolas $x^2 - y^2 = 1$ and $x^2 - y^2 = 4$ and on the other two sides by the ellipses $x^2/4 + y^2 = 1$ and $x^2/4 + y^2 = 4$.

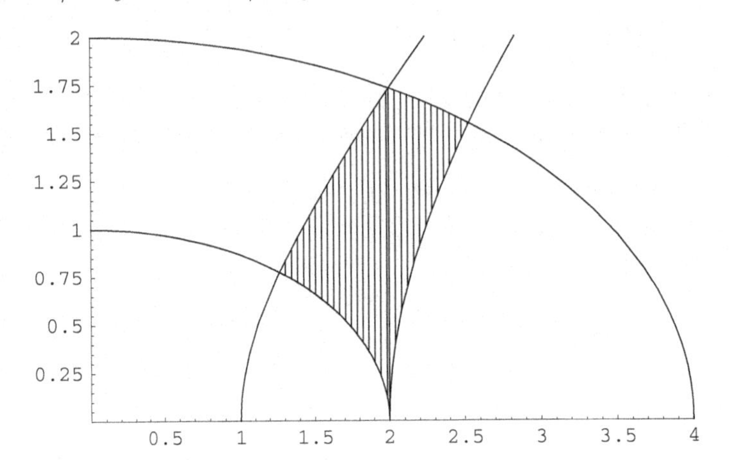

We, therefore, make the change of variables $u = x^2 - y^2$ and $v = x^2/4 + y^2$. Then

$$\frac{\partial(u,v)}{\partial(x,y)} = \begin{vmatrix} 2x & -2y \\ x/2 & 2y \end{vmatrix} = 5xy \quad \text{so} \quad \frac{\partial(x,y)}{\partial(u,v)} = 1/(5xy).$$

The integral greatly simplifies:

$$\iint_D \frac{xy}{y^2 - x^2}\, dA = \int_1^4 \int_1^4 \left(\frac{xy}{-u}\frac{1}{5xy}\right) du\, dv = -\frac{1}{5}\int_1^4 \int_1^4 \frac{1}{u}\, du\, dv$$
$$= -\frac{1}{5}\int_1^4 \ln 4\, dv = -\frac{3}{5}\ln 4.$$

19. (a) We are just generalizing what we've done to set up the area of a circle or the volume of a sphere (more recently see Exercises 11 and 14 from this section). Here our integral is:

$$\int_{-a}^{a} \int_{-\sqrt{a^2-x^2}}^{\sqrt{a^2-x^2}} \int_{-\sqrt{a^2-x^2-y^2}}^{\sqrt{a^2-x^2-y^2}} \int_{-\sqrt{a^2-x^2-y^2-z^2}}^{\sqrt{a^2-x^2-y^2-z^2}} dw\, dz\, dy\, dx.$$

(b) You should get $\pi^2 a^2/4$.

(c) For $n = 5$ you should get $8\pi^2 a^5/15$, and for $n = 6$ you should get $\pi^3 a^6/6$. If you include the cases for $n = 2$ and $n = 3$ you may begin to see a pattern for the even exponents. If n is even, the volume of the n-sphere of radius a is $\pi^{n/2}a^n/(n/2)!$. Fitting in the odd terms looks really hard and the pattern shouldn't occur to any of your students. In fact, the general formula depends on the Gamma function which is beyond what we would expect the students to know at this point. For kicks, the volume of the n-sphere of radius a is

$$\frac{\pi^{n/2}a^n}{\Gamma((n/2)+1)}.$$

Note that the volume of an n-sphere of radius a decreases to 0 as n increases.

23. (a) The one-dimensional ball of radius a consists of points in $\mathbf{R}$ described as

$$\{x_1 \in \mathbf{R} \mid x_1^2 \le a^2\} = \{x_1 \in \mathbf{R} \mid -a \le x_1 \le a\} = [-a, a].$$

The one-dimensional volume of this "ball" is the length of the interval; thus $V_1(a) = 2a$. The two-dimensional ball of radius a consists of points $(x_1, x_2) \in \mathbf{R}^2$ such that $x_1^2 + x_2^2 \le a^2$. Such points form a disk of radius a, so the two-dimensional volume of this disk is its area; hence $V_2(a) = \pi a^2$.

(b) By repeatedly using the recursive formula in Exercise 22, we have

$$V_n(a) = \begin{cases} \left(\frac{2\pi}{n} a^2\right)\left(\frac{2\pi}{n-2} a^2\right)\cdots\left(\frac{2\pi}{4} a^2\right) V_2(a) & \text{if } n \text{ is even,} \\ \left(\frac{2\pi}{n} a^2\right)\left(\frac{2\pi}{n-2} a^2\right)\cdots\left(\frac{2\pi}{1} a^2\right) V_1(a) & \text{if } n \text{ is odd.} \end{cases}$$

In the expressions for $V_n(a)$ above, there are $\frac{n}{2} - 1$ factors appearing before $V_2(a)$ when n is even and $\frac{n-1}{2}$ factors appearing before $V_1(a)$ when n is odd. Hence, using the results of part (a), we have

$$V_n(a) = \begin{cases} \dfrac{2^{(n/2)-1}\pi^{(n/2)-1}\left(a^2\right)^{(n/2)-1}\pi a^2}{n(n-2)\cdots 4} & \text{if } n \text{ is even} \\ \dfrac{2^{(n-1)/2}\pi^{(n-1)/2}\left(a^2\right)^{(n-1)/2} 2a}{n!!} & \text{if } n \text{ is odd} \end{cases}$$

$$= \begin{cases} \dfrac{2^{n/2}\pi^{n/2}\left(a^2\right)^{n/2}}{n(n-2)\cdots 4\cdot 2} & \text{if } n \text{ is even} \\ \dfrac{2^{(n+1)/2}\pi^{(n-1)/2}\left(a^2\right)^{(n-1)/2} a}{n!!} & \text{if } n \text{ is odd} \end{cases}$$

$$= \begin{cases} \dfrac{2^{n/2}\pi^{n/2}a^n}{2(n/2)\cdot 2\left((n/2)-1\right)\cdot 2\left((n/2)-2\right)\cdots(2\cdot 1)} & \text{if } n \text{ is even} \\ \dfrac{2^{(n+1)/2}\pi^{(n-1)/2}a^n}{n!!} & \text{if } n \text{ is odd} \end{cases}$$

$$= \begin{cases} \dfrac{\pi^{n/2}a^n}{(n/2)!} & \text{if } n \text{ is even} \\ \dfrac{2^{(n+1)/2}\pi^{(n-1)/2}a^n}{n!!} & \text{if } n \text{ is odd.} \end{cases}$$

27. **(a)**

$$I(\epsilon,\delta) = \int_\epsilon^{1-\epsilon}\int_\delta^{1-\delta} \frac{1}{\sqrt{xy}}\,dy\,dx = \int_\epsilon^{1-\epsilon} \frac{2}{\sqrt{x}}(\sqrt{1-\delta} - \sqrt{\delta})\,dx$$
$$= 4(\sqrt{1-\epsilon} - \sqrt{\epsilon})(\sqrt{1-\delta} - \sqrt{\delta})$$

(b) $\lim\limits_{(\epsilon,\delta)\to(0,0)} I(\epsilon,\delta) = 4\cdot 1\cdot 1 = 4$

31. Define $B_\epsilon = \{(x,y,z) | \epsilon \le x^2 + y^2 + z^2 \le 1\}$. Then

$$\iiint_B \ln\sqrt{x^2+y^2+z^2}\,dV = \lim_{\epsilon\to 0}\iiint_{B_\epsilon} \ln\sqrt{x^2+y^2+z^2}\,dV$$
$$= \lim_{\epsilon\to 0}\int_0^{2\pi}\int_\epsilon^1\int_0^\pi ((\ln\rho)\rho^2\sin\varphi)\,d\varphi\,d\rho\,d\theta$$
$$= \lim_{\epsilon\to 0}\int_0^{2\pi}\int_\epsilon^1 ((2\rho^2\ln\rho))\,d\rho\,d\theta$$
$$= \lim_{\epsilon\to 0}\int_0^{2\pi} ((2/3)\rho^3\ln\rho - 2\rho^3/9)\Big|_\epsilon^1\,d\theta$$
$$= \lim_{\epsilon\to 0} 2\pi\left(-\frac{2}{9} - \frac{2\epsilon^3}{3}\ln\epsilon + \frac{2\epsilon^3}{9}\right) = -4\pi/9.$$

37. **(a)** $\int_{-1}^{1} e^{-x^2}\,dx$ is finite since e^{-x^2} is bounded on $[-1,1]$. Since $0 \le e^{-x^2} \le 1/x^2$ on both $[1,\infty)$ and $(-\infty,-1]$ and the improper integrals $\int_1^{\infty}(1/x^2)\,dx$ and $\int_{-\infty}^{-1}(1/x^2)\,dx$ converge, we see that $\int_1^{\infty} e^{-x^2}\,dx$ and $\int_{-\infty}^{-1} e^{-x^2}\,dx$ both converge. Hence $\int_{-\infty}^{\infty} e^{-x^2}\,dx = \int_{-\infty}^{-1} e^{-x^2}\,dx + \int_{-1}^{1} e^{-x^2}\,dx + \int_{1}^{\infty} e^{-x^2}\,dx$ converges.

(b) We have

$$I^2 = \left(\int_{-\infty}^{\infty} e^{-x^2}\,dx\right)\left(\int_{-\infty}^{\infty} e^{-x^2}\,dx\right) = \left(\int_{-\infty}^{\infty} e^{-x^2}\,dx\right)\left(\int_{-\infty}^{\infty} e^{-y^2}\,dy\right)$$
$$= \int_{-\infty}^{\infty}\int_{-\infty}^{\infty} e^{-x^2}e^{-y^2}\,dx\,dy = \iint_{\mathbf{R}^2} e^{-x^2-y^2}\,dA.$$

(c) We'll use polar coordinates.

$$\iint_{D_a} e^{-x^2-y^2}\,dA = \int_0^{2\pi}\int_0^{a} e^{-r^2}\cdot r\,dr\,d\theta = -\frac{1}{2}\int_0^{2\pi}(e^{-a^2}-1)\,d\theta$$
$$= \pi(1-e^{-a^2})$$

(d) Note that, as $a \to \infty$, the disk D_a fills out more and more of $\mathbf{R}^2$. Thus $\lim_{a\to\infty}\iint_{D_a} e^{-x^2-y^2}\,dA = \iint_{\mathbf{R}^2} e^{-x^2-y^2}\,dA = I^2$.

(e) Putting everything together:

$$I^2 = \lim_{a\to\infty}\pi(1-e^{-a^2}) = \pi.$$

Thus $I = \int_{-\infty}^{\infty} e^{-x^2}\,dx = \sqrt{\pi}$.

41. First, we know that $C \ge 0$. Second,

$$1 = \int_{-\infty}^{\infty}\int_{-\infty}^{\infty} Ce^{-a|x|-b|y|}\,dx\,dy = 4\int_0^{\infty}\int_0^{\infty} Ce^{-ax-by}\,dx\,dy$$
$$= 4C\left[\int_0^{\infty} e^{-ax}\,dx\right]\left[\int_0^{\infty} e^{-by}\,dy\right] = 4C\left[-\frac{1}{a}e^{-ax}\Big|_0^{\infty}\right]\left[-\frac{1}{b}e^{-by}\Big|_0^{\infty}\right] = \frac{4C}{ab}.$$

So $C = ab/4$.

45. We use polar coordinates:

$$\int_{-1/2}^{1/2}\int_{-\sqrt{(1/4)-x^2}}^{\sqrt{(1/4)-x^2}} \frac{1}{\pi}e^{-x^2-y^2}\,dy\,dx = \int_0^{2\pi}\int_0^{1/2}\left(\frac{1}{\pi}re^{-r^2}\right)dr\,d\theta$$
$$= \int_0^{2\pi}\left(\frac{1}{\pi}\left(\frac{1}{2}\right)(1-e^{-1/4})\right)d\theta$$
$$= 1-e^{-1/4} \approx .22199.$$

Chapter 6

Line Integrals

6.1 Scalar and Vector Line Integrals

3.

$$\int_{\mathbf{x}} f\,ds = \int_1^3 \left[\frac{t+t^{3/2}}{t+t^{3/2}}\sqrt{1+1+\frac{9}{4}t}\right] dt = \int_1^3 \sqrt{2+\frac{9}{4}t}\,dt = \frac{8}{27}\left(2+\frac{9}{4}t\right)^{3/2}\Bigg|_1^3$$
$$= (35\sqrt{35} - 17\sqrt{17})/27.$$

7.

$$\int_{\mathbf{x}} f\,ds = \int_0^1 [(2t-t)\sqrt{1+4t^2}]\,dt + \int_1^3 (2-1+2t^2-4t+2)\,dt$$
$$= (5^{3/2}-1)/12 + 22/3 = (5^{3/2}+87)/12.$$

11.

$$\int_{\mathbf{x}} \mathbf{F}\cdot d\mathbf{s} = \int_{-1}^1 (t^3-t^2, t^{11})\cdot(2t, 3t^2)\,dt$$
$$= \int_{-1}^1 (2t^4 - 2t^3 + 3t^{13})\,dt = \left(\frac{2}{5}t^5 - \frac{1}{2}t^4 + \frac{3}{14}t^{14}\right)\Bigg|_{-1}^1 = \frac{4}{5}.$$

15.

$$\int_{\mathbf{x}} \mathbf{F}\cdot d\mathbf{s} = \int_0^{4\pi} (t, \sin^2 t, 2t)\cdot(-\sin t, \cos t, 1/3)\,dt = \int_0^{4\pi} (-t\sin t + \sin^2 t\cos t + 2t/3)\,dt$$
$$= \left(t\cos t - \sin t + \frac{\sin^3 t}{3} + \frac{t^2}{3}\right)\Bigg|_0^{4\pi} = \frac{12\pi + 16\pi^2}{3}.$$

19. The good news is: there is a ton of cancellation.

$$\int_{\mathbf{x}} \frac{x\,dx + y\,dx}{(x^2+y^2)^{3/2}} = \int_0^{2\pi} \frac{e^{2t}\cos 3t(2e^{2t}\cos 3t - 3e^{2t}\sin 3t) + e^{2t}\sin 3t(2e^{2t}\sin 3t + 3e^{2t}\cos 3t)}{(e^{4t}\cos^2 3t + e^{4t}\sin^2 3t)^{3/2}}\,dt$$
$$= \int_0^{2\pi} 2e^{-2t}\,dt = -e^{-2t}\Big|_0^{2\pi} = 1 - e^{-4\pi}.$$

23. First we organize the information we need for each of the four paths (each is for $0 \le t \le 1$).

i	$\mathbf{x}_i$	$\mathbf{x}_i'(t)$	$\mathbf{F}(\mathbf{x}_i(t))\cdot\mathbf{x}_i'(t)$
1	$(1-2t, 1, 3)$	$(-2, 0, 0)$	$-2(486 - 3(1-2t))$
2	$(-1, 1-2t, 3)$	$(0, -2, 0)$	$-2(-1)$
3	$(-1+2t, -1, 3)$	$(2, 0, 0)$	$2(486 - 3(1-2t))$
4	$(1, -1+2t, 3)$	$(0, 2, 0)$	$2(-1)$

So

$$\begin{aligned}\text{Work} &= \int_C \mathbf{F}\cdot d\mathbf{s} = \sum_{i=1}^{4}\int_{\mathbf{x}_i}\mathbf{F}(\mathbf{x}_i(t))\cdot\mathbf{x}_i'(t)\,dt \\ &= \int_0^1 (-2(486-3(1-2t)))\,dt + \int_0^1 (2)\,dt + \int_0^1 (2(486-3(1-2t)))\,dt + \int_0^1 (-2)\,dt \\ &= 0.\end{aligned}$$

27. Parametrize C as $\mathbf{x}(t) = (3-t,(3-t)^2)$, $0 \le t \le 3$, so that the parabola is oriented correctly. Then

$$\begin{aligned}\int_C y\,dx - x\,dy &= \int_0^3 \left[(3-t)^2(-1) - (3-t)(-2(3-t))\right]\,dt \\ &= \int_0^3 (3-t)^2\,dt = -\tfrac{1}{3}(3-t)^3\Big|_0^3 = 9.\end{aligned}$$

31. The path is $\mathbf{x}(t) = (4t+1, 2t+1, -t+2)$, $0 \le t \le 1$. The integral is

$$\begin{aligned}\int_C yz\,dx - xz\,dy + xy\,dz &= \int_0^1 [4(-2t^2+3t+2) - 2(-4t^2+7t+2) - (8t^2+6t+1)]\,dt \\ &= \int_0^1 [-8t^2-8t+3]\,dt = -\frac{11}{3}.\end{aligned}$$

35. **(a)** The force that Sisyphus is applying is $50\mathbf{x}'(t)/\|\mathbf{x}'(t)\|$. The path is given as $\mathbf{x}(t) = (5\cos 3t, 5\sin 3t, 10t)$ and so $\mathbf{x}'(t) = (-15\sin 3t, 15\cos 3t, 10)$ and $\|\mathbf{x}'(t)\| = \sqrt{325}$. The total work done is

$$\int_0^{10} \frac{50\mathbf{x}'(t)}{\|\mathbf{x}'(t)\|}\cdot\mathbf{x}'(t)\,dt = \int_0^{10} 50\|\mathbf{x}'(t)\|\,dt = \int_0^{10} 50\sqrt{325}\,dt = 2500\sqrt{13}\text{ ft-lb}.$$

(b) This time the 75 pounds is applied straight down. The total work done is

$$\int_0^{10} (0,0,75)\cdot(-15\sin 3t, 15\cos 3t, 10)\,dt = \int_0^{10} 750\,dt = 7500\text{ ft-lb}.$$

39. If $\mathbf{x}$ is a parametrization of C, then formula (3) of §6.1 gives $\int_C \nabla f\cdot d\mathbf{s} = \int_{\mathbf{x}} (\nabla f\cdot\mathbf{T})\,ds$, where $\mathbf{T} = \mathbf{x}'(t)/\|\mathbf{x}'(t)\|$. But $\mathbf{T}$ is tangent to C and ∇f is perpendicular to level sets of f (including C), so $\nabla f\cdot\mathbf{T} = 0$, and thus the integral must be zero.

43. **(a)** We have $\mathbf{x}_0 = (0,0,0)$, $\mathbf{x}_1 = \left(\frac{1}{4},\frac{1}{2},\frac{3}{4}\right)$, $\mathbf{x}_2 = \left(\frac{1}{2},1,\frac{3}{2}\right)$, $\mathbf{x}_3 = \left(\frac{3}{4},\frac{3}{2},\frac{9}{4}\right)$, $\mathbf{x}_4 = (1,2,3)$. Then all $\Delta x_k = \frac{1}{4}$, $\Delta y_k = \frac{1}{2}$, $\Delta z_k = \frac{3}{4}$. Then

$$\begin{aligned}T_4 &= \left(0+2\cdot\frac{3}{8}+2\cdot\frac{3}{2}+2\cdot\frac{27}{8}+6\right)\frac{1/4}{2} + (0+2\cdot 1+2\cdot 2+2\cdot 3+4)\frac{1/2}{2} \\ &+ \left(0+2\cdot\frac{1}{32}+2\cdot\frac{1}{4}+2\cdot\frac{27}{32}+2\right)\frac{3/4}{2} = \frac{245}{32} = 7.65625.\end{aligned}$$

(b) Parametrize C as $\begin{cases} x = t \\ y = 2t, \quad 0 \le t \le 1. \\ z = 3t \end{cases}$ Then

$$\begin{aligned}\int_C yz\,dx + (x+z)\,dy + x^2y\,dz &= \int_0^1 (6t^2 + 4t\cdot 2 + 2t^3\cdot 3)\,dt \\ &= \left(2t^3 + 4t^2 + \frac{3}{2}t^4\right)\Big|_0^1 = \frac{15}{2} = 7.5.\end{aligned}$$

6.2 Green's Theorem

3. $M(x, y) = y$ and $N(x, y) = x^2$.

- For the line integral, the path is split into four pieces, in each case $0 \le t \le 1$: $\mathbf{x}_1(t) = (1-2t, 1)$, $\mathbf{x}_2(t) = (-1, 1-2t)$, $\mathbf{x}_3(t) = (-1+2t, -1)$, and $\mathbf{x}_4(t) = (1, -1+2t)$.

$$\begin{aligned}\oint_{\partial D} M\,dx + N\,dy &= \int_0^1 [-2(1) + -2(1) + 2(-1) + 2(1)]\,dt \\ &= \int_0^1 -4\,dt = -4.\end{aligned}$$

- The area calculation is straightforward:

$$\iint_D (N_x - M_y)\,dA = \iint_D (2x-1)\,dA = \int_{-1}^1 \int_{-1}^1 (2x-1)\,dx\,dy = \int_{-1}^1 -2\,dy = -4.$$

9. Note that the curve is oriented clockwise so the square lies on the right side of the curve.

$$\oint_C (x^2 - y^2)\,dx + (x^2 + y^2)\,dy = -\int_0^1 \int_0^1 (2x + 2y)\,dy\,dx = -\int_0^1 (2x+1)\,dx = -2.$$

13. By Green's theorem:

$$\begin{aligned}\oint_C (x^4y^5 - 2y)\,dx + (3x + x^5y^4)\,dy &= -\iint_D ((3 + 5x^4y^4) - (5x^4y^4 - 2))\,dA \\ &= -\iint_D 5\,dA = -5 \cdot \text{area of } D = -5(2+3+4) = -45.\end{aligned}$$

(Note the minus sign because of the orientation of the curve.)

15. **(a)** Shown below are three views of the curve $\mathbf{x}(t) = (1 - t^2, t^3 - t)$.

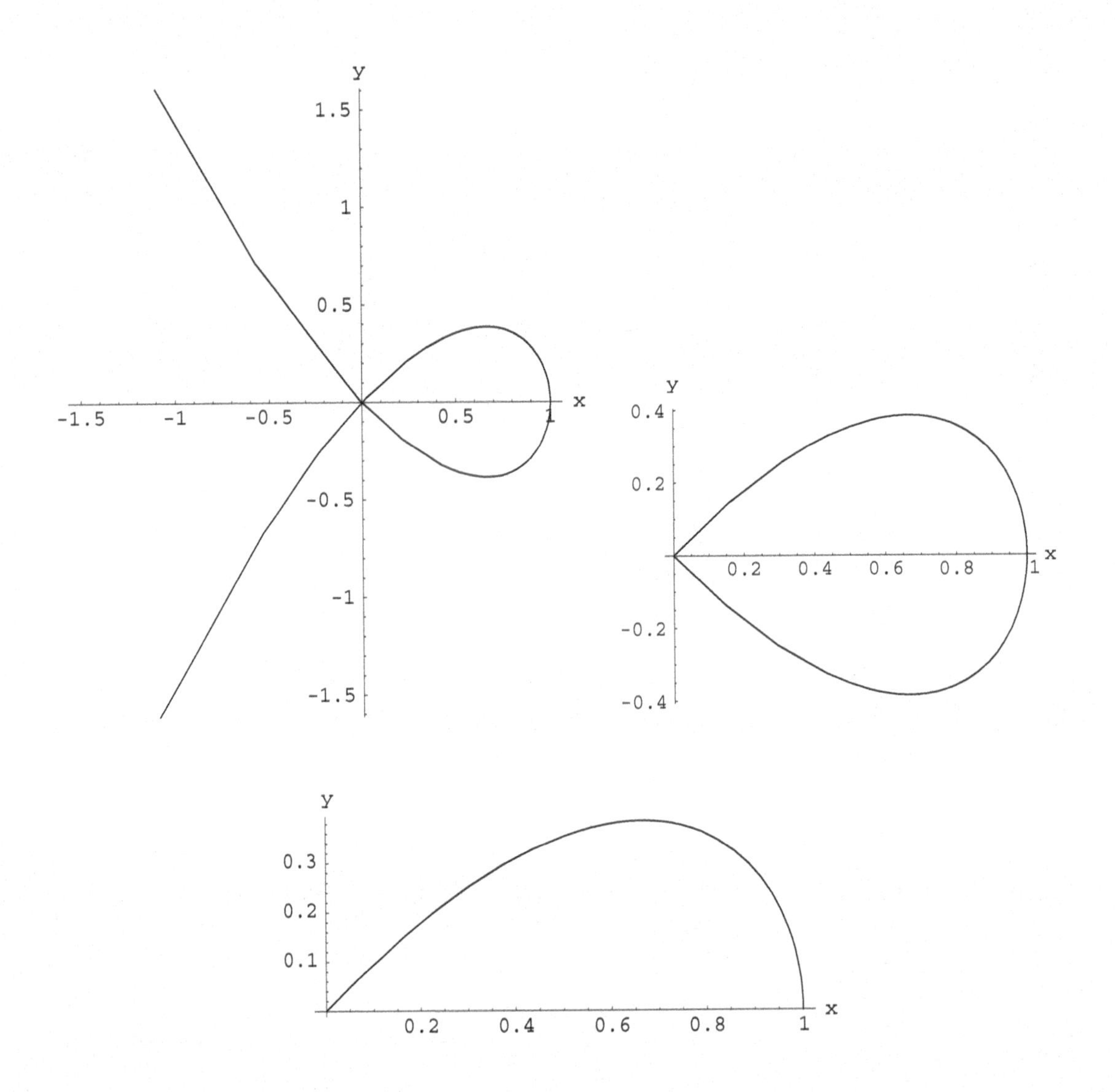

The figure on the top left is for $-3 \leq t \leq 3$, the top right figure is for $-1 \leq t \leq 1$, and the figure on the bottom is for $-1 \leq t \leq 0$. The first gives a feel for the curve, the second isolates the closed portion of the curve and the third gives us the orientation: that as t increases from -1 to 1 the path moves clockwise.

(b) We again must make an adjustment because the path moves clockwise. The area is

$$\frac{1}{2}\oint_{\partial D} -y\,dx + x\,dy = -\int_{-1}^{1} [(t^3 - t)(2t) + (1 - t^2)(3t^2 - 1)]\,dt$$

$$= \int_{-1}^{1} (t^4 - 2t^2 + 1)\,dt = \frac{8}{15}.$$

19. The area inside the polygon may be computed from

$$\frac{1}{2}\oint_C -y\,dx + x\,dy.$$

The key is to parametrize the boundary C of the polygon. This may be done in n line segment pieces. For $k = 1, \ldots, n-1$, the line segment from (a_k, b_k) to (a_{k+1}, b_{k+1}) is

$$\mathbf{x}_k(t) = ((a_{k+1} - a_k)t + a_k, (b_{k+1} - b_k)t + b_k), \quad 0 \le t \le 1,$$

while the last segment from (a_n, b_n) to (a_1, b_1) is

$$\mathbf{x}_n(t) = ((a_1 - a_n)t + a_n, (b_1 - b_n)t + b_n), \quad 0 \le t \le 1.$$

Thus, for $k = 1, \ldots, n-1$, we have

$$\begin{aligned}
&\frac{1}{2}\int_{\mathbf{x}_k} - y\,dx + x\,dy \\
&= \frac{1}{2}\int_0^1 [((b_k - b_{k+1})t - b_k)(a_{k+1} - a_k) + ((a_{k+1} - a_k)t + a_k)(b_{k+1} - b_k)]\,dt \\
&= \frac{1}{2}\int_0^1 [(b_k - b_{k+1})(a_{k+1} - a_k)t - b_k(a_{k+1} - a_k) \\
&\qquad + (a_{k+1} - a_k)(b_{k+1} - b_k)t + a_k(b_{k+1} - b_k)]\,dt \\
&= \frac{1}{2}\int_0^1 (-a_{k+1}b_k + a_kb_k + a_kb_{k+1} - a_kb_k)\,dt \\
&= \frac{1}{2}\int_0^1 (-a_{k+1}b_k + a_kb_{k+1})\,dt = \frac{1}{2}(-a_{k+1}b_k + a_kb_{k+1}) = \frac{1}{2}\begin{vmatrix} a_k & b_k \\ a_{k+1} & b_{k+1} \end{vmatrix}.
\end{aligned}$$

For the last segment, the calculation is very similar, so we abbreviate the steps:

$$\begin{aligned}
&\frac{1}{2}\int_{\mathbf{x}_k} - y\,dx + x\,dy \\
&= \frac{1}{2}\int_0^1 [((b_n - b_1)t - b_n)(a_1 - a_n) + ((a_1 - a_n)t + a_n)(b_1 - b_n)]\,dt \\
&= \frac{1}{2}\int_0^1 (-a_1b_n + a_nb_1)\,dt = \frac{1}{2}(-a_1b_n + a_nb_1) = \frac{1}{2}\begin{vmatrix} a_n & b_n \\ a_1 & b_1 \end{vmatrix}.
\end{aligned}$$

Adding the results of these calculations, we obtain

$$\frac{1}{2}\oint_C -y\,dx + x\,dy = \frac{1}{2}\left(\begin{vmatrix} a_1 & b_1 \\ a_2 & b_2 \end{vmatrix} + \begin{vmatrix} a_2 & b_2 \\ a_3 & b_3 \end{vmatrix} + \cdots + \begin{vmatrix} a_{n-1} & b_{n-1} \\ a_n & b_n \end{vmatrix} + \begin{vmatrix} a_n & b_n \\ a_1 & b_1 \end{vmatrix}\right),$$

as desired.

23. **(a)** By the divergence theorem:

$$\oint_C (2y\mathbf{i} - 3x\mathbf{j}) \cdot \mathbf{n}\,ds = \iint_D [(2y)_x + (-3x)_y]\,dA = \iint_D 0\,dA = 0.$$

(b) For direct computation, $\mathbf{n} = (\cos\theta, \sin\theta)$ and $x = \cos\theta$ and $y = \sin\theta$. Therefore,

$$\mathbf{F} \cdot \mathbf{n} = (2y, -3x) \cdot (\cos\theta, \sin\theta) = 2\cos\theta\sin\theta - 3\cos\theta\sin\theta = -\cos\theta\sin\theta = -\frac{1}{2}\sin 2\theta.$$

Then

$$\oint_C \mathbf{F} \cdot \mathbf{n}\,ds = -\frac{1}{2}\int_0^{2\pi} \sin 2\theta\,d\theta = 0.$$

27. Let $\delta = 1$ if C is oriented counterclockwise and $\delta = -1$ if C is oriented clockwise. Let D be the region bounded by C. Then by Green's Theorem,

$$\oint_C (x^2y^3 - 3y)\,dx + x^3y^2\,dy = \delta\iint_D (3x^2y^2 - 3x^2y^2 + 3)\,dA = 3\delta \text{ (the area of the rectangle)}.$$

31. First, $\oint_{\partial D} \frac{\partial f}{\partial n}\, ds = \oint_{\partial D} \nabla f \cdot \mathbf{n}\, ds = \oint_{\partial D} \left(\frac{\partial f}{\partial x}\mathbf{i} + \frac{\partial f}{\partial y}\mathbf{j}\right) \cdot \mathbf{n}\, ds$. You can continue the calculation or note that this is the same computation done in the proof of the divergence theorem with $\mathbf{F} = M\mathbf{i} + N\mathbf{j} = \frac{\partial f}{\partial x}\mathbf{i} + \frac{\partial f}{\partial y}\mathbf{j}$. Therefore, applying Green's theorem,

$$\oint_{\partial D} \frac{\partial f}{\partial n}\, ds = \oint_{\partial D} \frac{\partial f}{\partial y}\, dx + \frac{\partial f}{\partial x}\, dy = \iint_D \left(\frac{\partial^2 f}{\partial x^2} + \frac{\partial^2 f}{\partial y^2}\right) dA = \iint_D \nabla^2 f\, dA.$$

6.3 Conservative Vector Fields

1. **(a)** Let C be the path parametrized by $\mathbf{x}(t) = (t, t, t)$ with $0 \le t \le 1$. Then

$$\int_C z^2\, dx + 2y\, dy + xz\, dz = \int_0^1 (t^2 + 2t + t^2)\, dt = 2\int_0^1 (t^2 + t)\, dt = \frac{5}{3}.$$

(b) Let $\mathbf{x}(t) = (t, t^2, t^3)$ with $0 \le t \le 1$. Then

$$\int_C z^2\, dx + 2y\, dy + xz\, dz = \int_0^1 (t^6 + 2t^2(2t) + t^4(3t^2))\, dt = 4\int_0^1 (t^6 + t^3)\, dt = \frac{11}{7}.$$

(c) Parts (a) and (b) show that line integrals are not path-independent. By Theorem 3.3, therefore, $\mathbf{F}$ is not conservative.

5. $\frac{\partial N}{\partial x} = 3x^2 \sin y + \frac{1 - x^2y^2}{(1 + x^2y^2)^2} \neq -3x^2 \sin y + \frac{1 - x^2y^2}{(1 + x^2y^2)^2} = \frac{\partial M}{\partial y}$, so $\mathbf{F}$ is not conservative.

9. $\frac{\partial N}{\partial x} = 12xy = \frac{\partial M}{\partial y}$, so $\mathbf{F}$ is conservative. $\mathbf{F} = \nabla\left(3x^2y^2 - x^3 + \frac{1}{3}y^3\right)$.

In Exercises 13 and 17, we will check to see whether $\mathbf{F} = M\mathbf{i} + N\mathbf{j} + P\mathbf{k}$ is conservative by checking whether $\nabla \times \mathbf{F} = \mathbf{0}$. This amounts to checking whether $\partial N/\partial x = \partial M/\partial y, \partial P/\partial x = \partial M/\partial z$, and $\partial P/\partial y = \partial N/\partial z$. We also need to check that the domain of $\mathbf{F}$ is simply-connected. This last condition is only an issue in Exercise 16.

13. We see that $\partial N/\partial x = 1 = \partial M/\partial y$, $\partial P/\partial x = 0 = \partial M/\partial z$, and $\partial P/\partial y = \cos yz - yz \sin yz = \partial N/\partial z$. So $\mathbf{F}$ is conservative. $\mathbf{F} = \nabla(x^2 + xy + \sin yz)$.

17. We see that $\partial N/\partial x = ze^{-yz} + e^{xyz}(xyz^2 + z) = -\partial M/\partial y$, so $\mathbf{F}$ is not conservative.

21. For $\mathbf{F}$ to be conservative, we must have $\frac{\partial N}{\partial x} = \frac{\partial}{\partial y}(ye^{2x} + 3x^2e^y) = e^{2x} + 3x^2e^y$. Thus $N(x, y) = \frac{1}{2}e^{2x} + x^3e^y + u(y)$ where u is any C^1 function of y.

27. Here

$$\frac{\partial}{\partial x}\left(\frac{y}{\sqrt{x^2 + y^2}}\right) = -\frac{xy}{(x^2 + y^2)^{3/2}} = \frac{\partial}{\partial y}\left(\frac{x}{\sqrt{x^2 + y^2}}\right).$$

So $\mathbf{F}$ is conservative so long as we restrict the domain. Our domain must be simply-connected and must contain the upper half of the circle of radius 2 centered at the origin. Our domain also must not contain the origin as $\mathbf{F}$ is not defined at the origin. We can choose, for example, the upper half disk of radius 3 centered at the origin minus the upper half disk of radius one centered at the origin. This "semi-annular" region meets all of our conditions. Therefore, the given integral is path independent. We'll integrate along the path $\mathbf{x}(t) = (2\cos t, 2\sin t)$, $0 \le t \le \pi$. The integral

$$\int_C \frac{x\, dy + y\, dx}{\sqrt{x^2 + y^2}} = \int_0^\pi \frac{4\cos^2 t - 4\sin^2 t}{4\cos^2 t + 4\sin^2 t}\, dt = \int_0^\pi \cos 2t\, dt = \left.\frac{\sin 2t}{2}\right|_0^\pi = 0.$$

Using Theorem 3.3, and the fact that $\mathbf{F} = \nabla f$ where $f(x, y) = \sqrt{x^2 + y^2}$,

$$\int_C \frac{x\, dy + y\, dx}{\sqrt{x^2 + y^2}} = f(-2, 0) - f(2, 0) = \sqrt{(-2)^2 + 0} - \sqrt{2^2 + 0} = 0.$$

31. Use the result of Theorem 3.3. A potential function for $\mathbf{F}$ is $f(x, y, z) = x^2yz - xy^2z^3$. Therefore, the work is

$$f(6, 4, 2) - f(1, 1, 1) = -480 - 0 = -480.$$

35. **(a)** $\mathbf{F}$ is conservative since $\mathbf{F} = \nabla f$ where $f(x, y, z) = \sin(x^2 + xz) + \cos(y + yz)$.

(b) Since we have a potential function,

$$\int_{\mathbf{x}} \mathbf{F} \cdot d\mathbf{s} = \int_{\mathbf{x}} \nabla f \cdot d\mathbf{s} = f(\mathbf{x}(1)) - f(\mathbf{x}(0))$$

$$= f(1, 1, \pi - 1) - f(0, 0, 0) = -2.$$

37. You could check that $\mathbf{F}$ is conservative by confirming that $\nabla \times \mathbf{F} = \mathbf{0}$ on any simply-connected region that misses the origin. It is, however, easy enough to find the scalar potential for $\mathbf{F}$ is $f(x, y, z) = GMm(x^2 + y^2 + z^2)^{-1/2}$. So the work done by $\mathbf{F}$ as a particle of mass m moves from $\mathbf{x}_0$ to $\mathbf{x}_1$ is

$$f(x_1, y_1, z_1) - f(x_0, y_0, z_0) = \frac{GMm}{\sqrt{x_1^2 + y_1^2 + z_1^2}} - \frac{GMm}{\sqrt{x_0^2 + y_0^2 + z_0^2}} = GMm\left(\frac{1}{\|\mathbf{x}_1\|} - \frac{1}{\|\mathbf{x}_0\|}\right).$$

True/False Exercises for Chapter 6

1. True.
3. False. (The integral is negative.)
5. False. (The integral is 0.)
7. False. (There is equality only up to sign.)
9. True.
11. True.
13. True.
15. False. (The line integral could be $\pm \int_C \|\mathbf{F}\|\, ds$, depending on whether $\mathbf{F}$ points in the same or the opposite direction as C.)
17. False. (Let $\mathbf{F} = y\mathbf{i} - x\mathbf{j}$ and consider Green's theorem.)
19. False. (Under appropriate conditions, the integral is $f(B) - f(A)$.)
21. True.
23. False. (For the vector field to be conservative, the line integral must be zero for *all* closed curves, not just a particular one.)
25. False. (The vector field $(e^x \cos y \sin z, e^x \sin y \sin z, e^x \cos y \cos z)$ is not conservative.)
27. False. (The domain is not simply-connected.)
29. False. (f is only defined up to a constant.)

Miscellaneous Exercises for Chapter 6

1. Partition the curve into n pieces each of length $\Delta s_k =$ (length of C)$/n$. The right side of the given formula is just our calculation of arclength for a rectifiable curve:

$$\frac{\int_C f\, ds}{\text{length of } C} = \frac{\int_C f\, ds}{\int_C ds}.$$

Now, if on the kth sub-interval we choose any $\mathbf{c}_k$, then on the interval $f(\mathbf{x}) \approx f(\mathbf{c}_k)$. Therefore,

$$\lim_{n\to\infty}\left(\frac{1}{n}\sum_{k=1}^{n} f(\mathbf{c}_k)\right) = \lim_{n\to\infty}\sum_{k=1}^{n}\frac{f(\mathbf{c}_k)}{n} = \frac{\lim_{n\to\infty}\sum_{k=1}^{n} f(\mathbf{c}_k)\Delta s_k}{\text{length of } C} = \frac{\int_C f\, ds}{\text{length of } C}.$$

For each value of n we are calculating an average of n values of f at points on the curve. As n grows large, if this limit exists, it is reasonable to define it as the **average value** of f along C.

3. We may parametrize the semicircle as $\mathbf{x}(t) = (a\cos t, a\sin t)$, where $0 \le t \le \pi$. Therefore, $\|\mathbf{x}'(t)\| = a$. The length of the semicircle is πa and so the average y-coordinate may be found by calculating

$$\frac{1}{\pi a}\int_0^{\pi} a\sin t \cdot a\, dt = \frac{a}{\pi}\int_0^{\pi} \sin t\, dt = \frac{2a}{\pi}.$$

7. Locate the wire in the first quadrant of the xy-plane. Then the center is at $\left(\frac{a}{\sqrt{2}}, \frac{a}{\sqrt{2}}\right)$ and $\delta(x, y, z) = \left(x - \frac{a}{\sqrt{2}}\right)^2 + \left(y - \frac{a}{\sqrt{2}}\right)^2$. From symmetry considerations, we must have $\bar{x} = \bar{y}$. Now parametrize the quarter circle as

$$\begin{cases} x = a\cos t \\ y = a\sin t, \quad 0 \le t \le \pi/2. \end{cases}$$

Then $\|\mathbf{x}'(t)\| = a$. We have

$$\begin{aligned} M &= \int_C \delta\, ds = \int_0^{\pi/2} \left(\left(a\cos t - \frac{a}{\sqrt{2}} \right)^2 + \left(a\sin t - \frac{a}{\sqrt{2}} \right)^2 \right) a\, dt \\ &= a^3 \int_0^{\pi/2} \left(\cos^2 t - \sqrt{2}\cos t + \frac{1}{2} + \sin^2 t - \sqrt{2}\sin t + \frac{1}{2} \right) dt \\ &= a^3 \int_0^{\pi/2} (2 - \sqrt{2}\cos t - \sqrt{2}\sin t)\, dt = (\pi - 2\sqrt{2})a^3 \end{aligned}$$

$$\begin{aligned} M_{yz} &= \int_C x\delta\, ds = \int_0^{\pi/2} a\cos t \left(\left(a\cos t - \frac{a}{\sqrt{2}} \right)^2 + \left(a\sin t - \frac{a}{\sqrt{2}} \right)^2 \right) \cdot a\, dt \\ &= a^4 \int_0^{\pi/2} \cos t(2 - \sqrt{2}\cos t - \sqrt{2}\sin t)\, dt = a^4 \int_0^{\pi/2} \left(2\cos t - \frac{\sqrt{2}}{2}(1 + \cos 2t) - \sqrt{2}\sin t\cos t \right) dt \\ &= a^4 \left(2\sin t - \frac{\sqrt{2}}{2}t - \frac{\sqrt{2}}{4}\sin 2t - \frac{\sqrt{2}}{2}\sin^2 t \right) \Big|_0^{\pi/2} \\ &= a^4 \left(2 - \frac{\sqrt{2}\pi}{4} - 0 - \frac{\sqrt{2}}{2} - 0 \right) \\ &= \left(\frac{8 - \sqrt{2}\pi - 2\sqrt{2}}{4} \right) a^4. \end{aligned}$$

Hence

$$\bar{x} = \bar{y} = \frac{(8 - \sqrt{2}\pi - 2\sqrt{2})a^4}{4} \Big/ (\pi - 2\sqrt{2})a^3 = \left(\frac{8 - \sqrt{2}\pi - 2\sqrt{2}}{4(\pi - 2\sqrt{2})} \right) a.$$

13. We may parametrize the segment as $\mathbf{x}(t) = (1 - t)(-1, 1, 2) + t(2, 2, 3)$, $0 \le t \le 1$, or $\mathbf{x}(t) = (3t - 1, t + 1, t + 2)$. Then $\|\mathbf{x}'\| = \sqrt{9 + 1 + 1} = \sqrt{11}$.

$$\begin{aligned} I_z &= \int_C (x^2 + y^2)\delta\, ds = \int_0^1 [(3t - 1)^2 + (t + 1)^2][1 + (t + 2)^2] \cdot \sqrt{11}\, dt \\ &= \sqrt{11} \int_0^1 (10t^4 + 36t^3 + 36t^2 - 12t + 10)\, dt = \sqrt{11}(2t^5 + 9t^4 + 12t^3 - 6t^2 + 10t) \Big|_0^1 \\ &= \sqrt{11}(2 + 9 + 12 - 6 + 10) = 27\sqrt{11} \end{aligned}$$

$$\begin{aligned} M &= \int_C \delta\, ds = \int_0^1 (1 + (t + 2)^2)\sqrt{11}\, dt = \sqrt{11} \left(t + \frac{1}{3}(t + 2)^3 \right) \Big|_0^1 \\ &= \sqrt{11} \left(1 + 9 - \frac{8}{3} \right) = \frac{22\sqrt{11}}{3}. \end{aligned}$$

Hence $r_z = \sqrt{I_z/M} = \sqrt{\frac{81}{22}} = \frac{9\sqrt{22}}{22}$.

15. **(a)** $\oint_C g(x, y)\, ds = \int_{\mathbf{x}(\theta)} g(\mathbf{x}(\theta))\|\mathbf{x}'(\theta)\|\, d\theta = \int_a^b g(f(\theta)\cos\theta, f(\theta)\sin\theta)\sqrt{(f(\theta))^2 + (f'(\theta))^2}\, d\theta$.

(b) We'll use the formula from part (a).

$$\begin{aligned} \int_C g\, ds &= \int_0^{2\pi} ([(e^{3\theta}\cos\theta)^2 + (e^{3\theta}\sin\theta)^2 - 2(e^{3\theta}\cos\theta)]\sqrt{e^{6\theta} + 9e^{6\theta}})\, d\theta \\ &= \int_0^{2\pi} ([e^{6\theta} - 2e^{3\theta}\cos\theta]\sqrt{10}e^{3\theta})\, d\theta \\ &= \frac{\sqrt{10}}{333}(37e^{18\pi} - 108e^{12\pi} + 71). \end{aligned}$$

In this text κ is always non-negative. In cases where the curvature is signed, differential geometers are often interested in the **total squared curvature**: $\int_C \kappa^2\, ds$.

19. We parameterize the ellipse by the path $\mathbf{x}(t) = (a\cos t, b\sin t, 0)$ for $0 \le t \le 2\pi$. Then, using Exercise 16,

$$K = \int_0^{2\pi} \frac{\|(-a\sin t, b\cos t, 0) \times (-a\cos t, -b\sin t, 0)\|}{\|(-a\sin t, b\cos t, 0)\|^2}\, dt = \int_0^{2\pi} \frac{\|(0,0,ab)\|}{a^2\sin^2 t + b^2\cos^2 t}\, dt$$
$$= \int_0^{2\pi} \frac{ab}{a^2\sin^2 t + b^2\cos^2 t}\, dt = 2\pi.$$

This verifies Fenchel's Theorem for the given ellipse. This final integral was calculated using *Mathematica*. With work it can also be done by hand.

21. The work done is

$$\int_{\mathbf{x}} \mathbf{F}\cdot d\mathbf{s} = \int_0^1 (\sin(t^3), \cos(-t^2), t^4)\cdot(3t^2, -2t, 1)\, dt = \int_0^1 (3t^2\sin(t^3) - 2t\cos(-t^2) + t^4)\, dt$$
$$= \left(-\cos(t^3) + \sin(-t^2) + t^5/5\right)\Big|_0^1 = 6/5 - \cos 1 - \sin 1.$$

25. We begin by applying Green's theorem (here D has constant density δ):

$$\frac{1}{2\cdot\text{area of } D}\oint_{\partial D} x^2\, dy = \frac{1}{2\cdot\text{area of } D}\iint_D 2x\, dx\, dy = \frac{\iint_D x\, dx\, dy}{\iint_D dx\, dy}$$
$$= \frac{\iint_D x\,\delta\, dx\, dy}{\iint_D \delta\, dx\, dy} = \bar{x}.$$

Similarly,

$$\frac{1}{\text{area of } D}\oint_{\partial D} xy\, dy = \frac{1}{\text{area of } D}\iint_D y\, dx\, dy = \frac{\iint_D y\, dx\, dy}{\iint_D dx\, dy}$$
$$= \frac{\iint_D y\,\delta\, dx\, dy}{\iint_D \delta\, dx\, dy} = \bar{y}.$$

For the second pair of formulas, we proceed in an entirely similar manner with Green's theorem.

$$-\frac{1}{\text{area of } D}\oint_{\partial D} xy\, dx = -\frac{1}{\text{area of } D}\iint_D -x\, dx\, dy$$
$$= \frac{1}{\text{area of } D}\iint_D x\, dx\, dy = \frac{\iint_D x\, dx\, dy}{\iint_D dx\, dy}$$
$$= \frac{\iint_D x\,\delta\, dx\, dy}{\iint_D \delta\, dx\, dy} = \bar{x}.$$

Also,

$$-\frac{1}{2\cdot\text{area of } D}\oint_{\partial D} y^2\, dx = -\frac{1}{2\cdot\text{area of } D}\iint_D -2y\, dx\, dy$$
$$= \frac{1}{\text{area of } D}\iint_D y\, dx\, dy = \frac{\iint_D y\,\delta\, dx\, dy}{\iint_D \delta\, dx\, dy} = \bar{y}.$$

29. Apply the results of Exercise 28 to both parts of the line integral.

$$\oint_C (f\nabla g - g\nabla f)\cdot \mathbf{n}\,ds = \oint_C f\nabla g\cdot \mathbf{n}\,ds - \oint_C g\nabla f\cdot \mathbf{n}\,ds$$
$$= \iint_D (f\nabla^2 g + \nabla f\cdot\nabla g)\,dA - \iint_D (g\nabla^2 f + \nabla g\cdot\nabla f)\,dA$$
$$= \iint_D (f\nabla^2 g - g\nabla^2 f)\,dA$$

35. (a) The boundary is in two pieces which we separately parametrize as $\mathbf{x}_1(\theta) = (\cos\theta, \sin\theta)$ and $\mathbf{x}_2(\theta) = (a\cos\theta, -a\sin\theta)$, each for $0 \le \theta \le 2\pi$. The line integral is then

$$\oint_C \frac{-y}{x^2+y^2}\,dx + \frac{x}{x^2+y^2}\,dy = \int_0^{2\pi} (\sin^2\theta + \cos^2\theta)\,d\theta + \int_0^{2\pi}\left(-\frac{a^2\sin^2\theta}{a^2} - \frac{a^2\cos^2\theta}{a^2}\right)d\theta$$
$$= \int_0^{2\pi} (1-1)\,d\theta = 0.$$

The double integral is

$$\iint_D \left[\frac{\partial}{\partial x}\left(\frac{x}{x^2+y^2}\right) - \frac{\partial}{\partial y}\left(\frac{-y}{x^2+y^2}\right)\right]dx\,dy = \iint_D \left[\frac{-x^2+y^2}{(x^2+y^2)^2} + \frac{-y^2+x^2}{(x^2+y^2)^2}\right]dx\,dy$$
$$= \iint_D 0\,dx\,dy = 0.$$

Thus the conclusion of Green's theorem holds for **F** in the given annular region.

(b) This time the line integral is only taken over the outer boundary and so

$$\oint_C \frac{-y}{x^2+y^2}\,dx + \frac{x}{x^2+y^2}\,dy = 2\pi.$$

The same cancellation takes place in the double integral as in part (a), so

$$\iint_D \left[\frac{\partial}{\partial x}\left(\frac{x}{x^2+y^2}\right) - \frac{\partial}{\partial y}\left(\frac{-y}{x^2+y^2}\right)\right]dx\,dy = 0.$$

The problem is that **F** is not defined at the origin.

(c) Let D be the region so that ∂D consists of the given curve C oriented counterclockwise and also the curve C_a, the circle of radius a centered at the origin oriented clockwise. Then **F** is defined everywhere in the region D. Green's theorem holds so

$$\oint_C \frac{-y}{x^2+y^2}\,dx + \frac{x}{x^2+y^2}\,dy + \oint_{C_a} \frac{-y}{x^2+y^2}\,dx + \frac{x}{x^2+y^2}\,dy = \oint_{C\cup C_a} \frac{-y}{x^2+y^2}\,dx + \frac{x}{x^2+y^2}\,dy$$
$$= \iint_D \left(\frac{\partial}{\partial x}\left(\frac{x}{x^2+y^2}\right) - \frac{\partial}{\partial y}\left(\frac{-y}{x^2+y^2}\right)\right)dx\,dy = 0, \quad \text{but}$$

$$\oint_{C_a} \frac{-y}{x^2+y^2}\,dx + \frac{x}{x^2+y^2}\,dy = -2\pi. \quad \text{Therefore,} \quad \oint_C \frac{-y}{x^2+y^2}\,dx + \frac{x}{x^2+y^2}\,dy = 2\pi.$$

37. (a) By the divergence theorem, the flux

$$\int_C \mathbf{F}\cdot\mathbf{n}\,ds = \int_0^1\int_0^5 \left(\frac{\partial}{\partial x}[e^y] + \frac{\partial}{\partial y}[x^4]\right)dy\,dx = 0.$$

(b) Again, by the divergence theorem, the flux

$$\int_C \mathbf{F}\cdot\mathbf{n}\,ds = \iint_D \left(\frac{\partial}{\partial x}[f(y)] + \frac{\partial}{\partial y}[f(x)]\right)dy\,dx = 0.$$

Chapter 7

Surface Integrals and Vector Analysis

7.1 Parametrized Surfaces

1. (a) To find a normal vector we calculate

$$\mathbf{T}_s(s,t) = (2s, 1, 2s) \quad \text{so} \quad \mathbf{T}_s(2,-1) = (4,1,4)$$
$$\mathbf{T}_t(s,t) = (-2t, 1, 3) \quad \text{so} \quad \mathbf{T}_t(2,-1) = (2,1,3).$$

Then a normal vector is

$$\mathbf{N}(2,-1) = \mathbf{T}_s(2,-1) \times \mathbf{T}_t(2,-1) = (-1,-4,2).$$

(b) We find an equation for the tangent plane using

$$0 = \mathbf{N}(2,-1) \cdot (\mathbf{x} - (3,1,1)) = (-1,-4,2) \cdot (\mathbf{x} - (3,1,1)) = -x + 3 - 4y + 4 + 2z - 2.$$

This is equivalent to $x + 4y - 2z = 5$.

5. (a) Using *Mathematica* and the command:

ParametricPlot3D[{s, s^2 + t, t^2}, {s, −2, 2}, {t, −2, 2}, AxesLabel → {x,y,z}],

we obtain the image

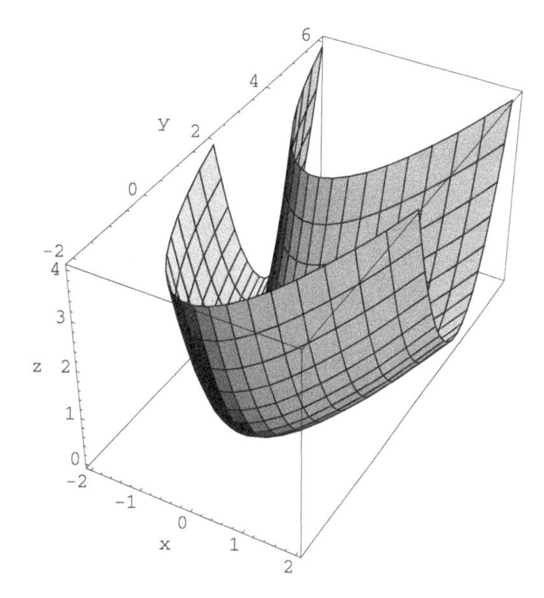

(b) To determine whether the surface is smooth we need to calculate $\mathbf{N}$. First, $\mathbf{T}_s(s,t) = (1, 2s, 0)$, and $\mathbf{T}_t(s,t) = (0,1,2t)$ so $\mathbf{N} = \mathbf{T}_s \times \mathbf{T}_t = (4st, -2t, 1)$. We conclude that $\mathbf{N} \neq \mathbf{0}$ for any (s, t) so $\mathbf{N}$ is smooth.

(c) If $(s, s^2 + t, t^2) = (1, 0, 1)$, then $s = 1$ and $t = -1$. So $\mathbf{N}(1,-1) = (-4, 2, 1)$ and an equation of the tangent plane at this point is $(-4,2,1) \cdot (\mathbf{x} - (1,0,1)) = 0$ or more simply, $4x - 2y - z = 3$.

11. (a) First we consider the sphere as the graph of the function $f(x,y) = \sqrt{4 - (x-2)^2 - (y+1)^2}$. The partial derivatives are

$$f_x = \frac{-(x-2)}{\sqrt{4-(x-2)^2-(y+1)^2}} \quad f_y = \frac{-(y+1)}{\sqrt{4-(x-2)^2-(y+1)^2}}.$$

So $f_x(1,0,\sqrt{2}) = 1/\sqrt{2}$ and $f_y(1,0,\sqrt{2}) = -1/\sqrt{2}$. By Theorem 3.3 of Chapter 2, $z = f(a,b) + f_x(a,b)(x-a) + f_y(a,b)(y-b)$. In this case, this is $z = \sqrt{2} + (1/\sqrt{2})(x-1) - (1/\sqrt{2})y$, or equivalently, $-x + y + \sqrt{2}z = 1$.

(b) Now we look at the sphere as a level surface of $F(x,y,z) = (x-2)^2 + (y+1)^2 + z^2$. The gradient $\nabla F(x,y,z) = 2(x-2, y+1, z)$ and, therefore, $\nabla F(1,0,\sqrt{2}) = (-2,2,2\sqrt{2})$. By formula (5) of Section 2.6, the tangent plane is given by

$$0 = \nabla F(1,0,\sqrt{2}) \cdot (\mathbf{x} - (1,0,\sqrt{2})) = (-2,2,2\sqrt{2}) \cdot (\mathbf{x} - (1,0,\sqrt{2})).$$

This too is equivalent to $-x + y + \sqrt{2}z = 1$.

(c) Now we'll use the results of this section. Considering the z-component, we see $2\cos s = \sqrt{2}$ so $\cos s = \sqrt{2}/2$. Considering the y- and x-components, $2\sin s \sin t = 1$ and $\sin s \cos t = -1$. Thus we have that $s = \pi/4$ and $t = 3\pi/4$. Also $\mathbf{T}_s(s,t) = (2\cos s\cos t, 2\cos s \sin t, -2\sin s)$ and $\mathbf{T}_t(s,t) = (-2\sin s \sin t, 2\sin s\cos t, 0)$. A normal vector to the sphere at the specified point is

$$\mathbf{N}(\pi/4, 3\pi/4) = \mathbf{T}_s(\pi/4, 3\pi/4) \times \mathbf{T}_t(\pi/4, 3\pi 4) = (-1,1,-\sqrt{2}) \times (-1,-1,0) = (-\sqrt{2}, \sqrt{2}, 2).$$

The tangent plane is given by $(-\sqrt{2},\sqrt{2},2)\cdot(\mathbf{x} - (1,0,\sqrt{2})) = 0$ which is also equivalent to $-x + y + \sqrt{2}z = 1$.

15. If we rewrite the equation for the hyperboloid as $z^2 = x^2 + y^2 + 1$, then we see that we must have $z = \pm\sqrt{x^2+y^2+1}$. Therefore, the hyperboloid may be parametrized with two maps as $\mathbf{X}_1\colon \mathbf{R}^2 \to \mathbf{R}^3$, $\mathbf{X}_1(s,t) = \left(s,t,\sqrt{s^2+t^2+1}\right)$ and $\mathbf{X}_2\colon \mathbf{R}^2 \to \mathbf{R}^3$, $\mathbf{X}_2(s,t) = \left(s,t,-\sqrt{s^2+t^2+1}\right)$.

19. (a) To find an equation for the tangent plane to a surface described by the equation $y = g(x,z)$ at the point $(a, g(a,c), c)$ we basically permute the case detailed in the text and in Exercise 18 to obtain either

$$(g_x(a,c), -1, g_z(a,c)) \cdot (\mathbf{x} - (a, g(a,c), c)) = 0 \quad \text{or}$$
$$g_x(a,c)(x-a) - (y - g(a,c)) + g_z(a,c)(z-c) = 0.$$

(b) Similarly, an equation for the tangent plane to a surface described by the equation $x = h(y,z)$ at the point $(h(b,c), b, c)$ is either

$$(-1, h_y(b,c), h_z(b,c)) \cdot (\mathbf{x} - (h(b,c), b, c)) = 0 \quad \text{or}$$
$$-(x - h(b,c)) + h_y(b,c)(y-b) + h_z(b,c)(z-c) = 0.$$

23. We need to calculate $\|\mathbf{T}_s(s,t) \times \mathbf{T}_t(s,t)\|$. We have that $\mathbf{X}(s,t) = (s+t, s-t, s)$ for $-1 \le s \le 1$ and $-\sqrt{1-s^2} \le t \le \sqrt{1-s^2}$. Therefore, $\mathbf{T}_s(s,t) = (1,1,1)$, $\mathbf{T}_t(s,t) = (1,-1,0)$, $\mathbf{T}_s(s,t)\times\mathbf{T}_t(s,t) = (1,1,-2)$ and $\|\mathbf{T}_s(s,t)\times\mathbf{T}_t(s,t)\| = \sqrt{6}$. So we are integrating $\sqrt{6}$ over the unit disk in the *st*-plane. Therefore, the surface area of $\mathbf{X}(D) = \iint_D \sqrt{6}\,dt\,ds = \sqrt{6}\pi$.

25. A quick look at the figure below shows a cutaway of a quarter of the *xz*-plane intersection of the cylindrical hole of radius b bored in a sphere of radius a. The height of the hole is $2\sqrt{a^2-b^2}$. The top half of the ring is the region swept out by the portion of the diagram containing the letter 'h'.

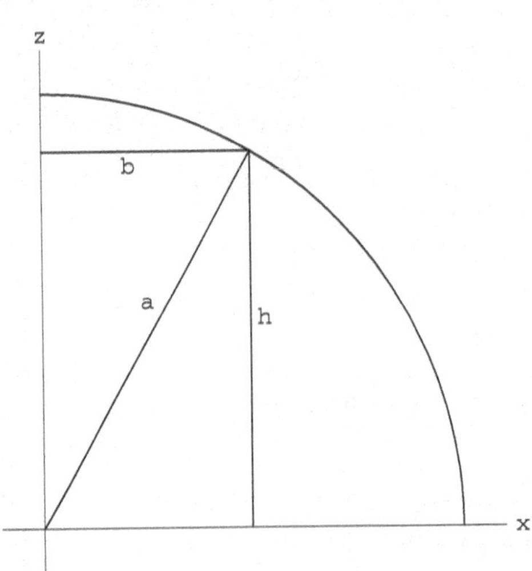

If $\mathbf{X}(s,t) = (a\sin s\cos t, a\sin s\sin t, a\cos s)$, then $\mathbf{T}_s(s,t) = (a\cos s\cos t, a\cos s\sin t, -a\sin s)$, $\mathbf{T}_t(s,t) = (-a\sin s\sin t, a\sin s\cos t, 0)$, $\mathbf{T}_s(s,t) \times \mathbf{T}_t(s,t) = a^2\sin s(\sin s\cos t, \sin s\sin t, \cos s)$ and $\|\mathbf{T}_s(s,t)\times\mathbf{T}_t(s,t)\| =$

$a^2 \sin s$. Notice that the angle s made with the z-axis has lower limit $\cos^{-1}(h/a) = \cos^{-1}(\sqrt{a^2 - b^2}/a)$ and upper limit $\pi/2$. So the surface area is

$$2\int_0^{2\pi} \int_{\cos^{-1}(\sqrt{a^2-b^2}/a)}^{\pi/2} a^2 \sin s \, ds \, dt = 2\int_0^{2\pi} a^2 \left(\frac{\sqrt{a^2 - b^2}}{a}\right) dt = 4\pi a\sqrt{a^2 - b^2}.$$

29. We have $z = f(x, y)$ and $f_x^2 + f_y^2 = a$ so, by formula (9), the surface area is

$$\iint_D \sqrt{f_x^2 + f_y^2 + 1}\, dx\, dy = \iint_D \sqrt{a+1}\, dx\, dy = \sqrt{a+1}(\text{area of } D).$$

33. We have $\begin{cases} x = f(\varphi, \theta) \sin\varphi \cos\theta \\ y = f(\varphi, \theta) \sin\varphi \sin\theta \\ z = f(\varphi, \theta) \cos\varphi \end{cases}$ from spherical/Cartesian conversions. From this,

$$\mathbf{T}_\varphi = (f_\varphi \sin\varphi \cos\theta + f \cos\varphi \cos\theta, f_\varphi \sin\varphi \sin\theta + f \cos\varphi \sin\theta, f_\varphi \cos\varphi - f \sin\varphi)$$
$$\mathbf{T}_\theta = (f_\theta \sin\varphi \cos\theta - f \sin\varphi \sin\theta, f_\theta \sin\varphi \sin\theta + f \sin\varphi \cos\theta, f_\theta \cos\varphi).$$

After some careful computation and using $\cos^2\alpha + \sin^2\alpha = 1$, we find $\mathbf{N}(\varphi, \theta) = \mathbf{T}_\varphi \times \mathbf{T}_\theta = (f f_\theta \sin\theta - f f_\varphi \sin\varphi \cos\varphi \cos\theta + f^2 \sin^2\varphi \cos\theta, f^2 \sin^2\varphi \sin\theta - f f_\varphi \sin\varphi \cos\varphi \sin\theta - f f_\theta \cos\theta, f^2 \sin\varphi \cos\varphi + f f_\varphi \sin^2\varphi)$. After still more computation, one finds $\|\mathbf{N}\|^2 = (f f_\theta)^2 + (f f_\varphi)^2 \sin^2\varphi + f^4 \sin^2\varphi$, so that using formula (6) in §7.1,

$$\begin{aligned}\text{Surface area} &= \iint_D \sqrt{(f f_\theta)^2 + (f f_\varphi)^2 \sin^2\varphi + f^4 \sin^2\varphi}\, d\varphi\, d\theta \\ &= \iint_D f(\varphi, \theta)\sqrt{f_\theta^2 + \sin^2\varphi(f_\varphi + f^2)}\, d\varphi\, d\theta.\end{aligned}$$

7.2 Surface Integrals

1. We will use Definition 2.1 to calculate the integral: $\iint_{\mathbf{X}} f\, dS = \iint_D f(\mathbf{X}(s,t)) \|\mathbf{T}_s \times \mathbf{T}_t\|\, ds\, dt$. Here $\mathbf{X}(s,t) = (s, s+t, t)$, $\mathbf{T}_s(s,t) = (1, 1, 0)$, $\mathbf{T}_t(s,t) = (0, 1, 1)$, $\mathbf{N}(s,t) = \mathbf{T}_s(s,t) \times \mathbf{T}_t(s,t) = (1, -1, 1)$, and $\|\mathbf{N}(s,t)\| = \sqrt{3}$. Also, $f(\mathbf{X}(s,t)) = s^2 + (s+t)^2 + t^2 = 2(s^2 + st + t^2)$. So

$$\begin{aligned}\iint_{\mathbf{X}} (x^2 + y^2 + z^2)\, dS &= 2\sqrt{3}\int_0^2 \int_0^1 (s^2 + st + t^2)\, ds\, dt = 2\sqrt{3}\int_0^2 \left(\frac{1}{3} + \frac{t}{2} + t^2\right) dt \\ &= 2\sqrt{3}\left(\frac{2}{3} + 1 + \frac{8}{3}\right) = \frac{26}{\sqrt{3}}.\end{aligned}$$

5. We will parametrize the six faces of the cube as follows (in each case $-2 \le s, t \le 2$):

i	$\mathbf{X}(s,t)$ for S_i	face
1	$(s, t, 2)$	top
2	$(s, t, -2)$	bottom
3	$(s, 2, t)$	right
4	$(s, -2, t)$	left
5	$(2, s, t)$	front
6	$(-2, s, t)$	back

Note that in each case $\|\mathbf{N}(s,t)\| = 1$, so $\iint_{S_i} [x(s,t)]^2 \|\mathbf{N}(s,t)\|\, ds\, dt = \iint_{S_i} [x(s,t)]^2 ds\, dt$ for $1 \le i \le 6$. Also,

$\iint_{S_i} [x(s,t)]^2\, ds\, dt = \int_{-2}^{2}\int_{-2}^{2} s^2\, ds\, dt$ for $i = 1, 2, 3, 4$ and $\iint_{S_i} [x(s,t)]^2\, ds\, dt = \int_{-2}^{2}\int_{-2}^{2} 4\, ds\, dt$ for $i = 5, 6$. Then

$$\begin{aligned}\iint_S x^2\, dS &= \sum_{i=1}^{6} \iint_{S_i} [x(s,t)]^2 \|\mathbf{N}(s,t)\|\, ds\, dt \\ &= 4\int_{-2}^{2}\int_{-2}^{2} s^2\, ds\, dt + 2\int_{-2}^{2}\int_{-2}^{2} 4\, ds\, dt \\ &= 4\int_{-2}^{2} \left.\frac{s^3}{3}\right|_{-2}^{2} dt + 8\int_{-2}^{2} s\big|_{-2}^{2}\, dt \\ &= 4\int_{-2}^{2} \frac{16}{3}\, dt + 8\int_{-2}^{2} 4\, dt = \frac{256}{3} + 128 = \frac{640}{3}.\end{aligned}$$

9. (a) We parametrize the cylinder as $\begin{cases} x = 2\cos t \\ y = 2\sin t \\ z = s \end{cases} \quad 0 \le t < 2\pi,\ -2 \le s \le 2.$

Then

$$\begin{aligned}\|\mathbf{T}_s \times \mathbf{T}_t\| &= \|(0,0,1) \times (-2\sin t, 2\cos t, 0)\| = \|(-2\cos t, -2\sin t, 0)\| \\ &= 2.\end{aligned}$$

Hence

$$\begin{aligned}\iint_S (z - x^2 - y^2)\, dS &= \int_0^{2\pi}\int_{-2}^{2} (s-4)\cdot 2\, ds\, dt = \int_0^{2\pi} \left.(s^2 - 8s)\right|_{s=-2}^{2} dt \\ &= \int_0^{2\pi} -32\, dt = -64\pi.\end{aligned}$$

(b) $\iint_S (z - x^2 - y^2)\, dS = \iint_S z\, dS - \iint_S (x^2+y^2)\, dS$. S is symmetric about the $z = 0$ plane and $x^2 + y^2 = 4$ on S. Hence $\iint_S z\, dS = 0$ and $-\iint_S (x^2+y^2)\, dS = -\iint_S 4\, dS = -4\cdot(\text{surface area of } S) = -4(4\pi\cdot 4) = -64\pi$.

13.

$$\begin{aligned}\iint_S x^2\, dS &= \int_0^4\int_0^{2\pi} 27\cos^2 s\, ds\, dt + 2\int_0^3\int_0^{2\pi} t^3\cos^2 s\, ds\, dt \\ &= 27\int_0^4 \left.\left[\frac{s}{2} + \frac{1}{9}\sin 2s\right]\right|_0^{2\pi} dt + 2\int_0^3 t^3 \left.\left[\frac{s}{2} + \frac{1}{4}\sin 2s\right]\right|_0^{2\pi} dt = 27\int_0^4 \pi\, dt + 2\int_0^3 \pi t^3\, dt \\ &= 108\pi + \frac{81\pi}{2} = \frac{297\pi}{2}.\end{aligned}$$

15.

$$\begin{aligned}\iint_S (z\mathbf{k})\cdot d\mathbf{S} &= \int_0^4\int_0^{2\pi} (0,0,t)\cdot(3\cos s, 3\sin s, 0)\, ds\, dt + \int_0^3\int_0^{2\pi} (0,0,0)\cdot(0,0,-t)\, ds\, dt \\ &\quad + \int_0^3\int_0^{2\pi} (0,0,4)\cdot(0,0,t)\, ds\, dt = \int_0^3\int_0^{2\pi} 4t\, ds\, dt = \int_0^3 8\pi t\, dt = 36\pi.\end{aligned}$$

A different approach would have been to notice that, since the unit normal vector to the lateral surface S_1 has no $\mathbf{k}$ component, $\iint_{S_1} z\mathbf{k}\cdot d\mathbf{S} = 0$. Also, $z = 0$ on S_2 so $\iint_{S_2} z\mathbf{k}\cdot d\mathbf{S} = 0$. Finally, $z = 4$ on S_3 and therefore

$$\iint_S z\mathbf{k}\cdot d\mathbf{S} = \iint_{S_3} z\mathbf{k}\cdot d\mathbf{S} = \iint_{S_3} 4\mathbf{k}\cdot\mathbf{k}\, dS = \iint_{S_3} 4\, dS = 4\cdot(\text{area of } S_3) = 4(\pi 3^2) = 36\pi.$$

19.

$$\iint_S (y\mathbf{j}) \cdot \mathbf{n}\, dS = \int_0^{2\pi} \int_0^{\pi/2} (0, a\sin\varphi \sin\theta, 0) \cdot (a^2 \sin^2 \varphi \cos\theta, a^2 \sin^2 \varphi \sin\theta, a^2 \cos\varphi \sin\varphi)\, d\varphi\, d\theta$$

$$= \int_0^{2\pi} \int_0^{\pi/2} (a^3 \sin^3 \varphi \sin^2 \theta)\, d\varphi\, d\theta = a^3 \int_0^{2\pi} \int_0^{\pi/2} (1 - \cos^2 \varphi) \sin\varphi \sin^2 \theta\, d\varphi\, d\theta$$

$$= a^3 \int_0^{2\pi} \left[-\cos\varphi + \frac{\cos^3 \varphi}{3} \right] \Bigg|_0^{\pi/2} \sin^2 \theta\, d\varphi\, d\theta = \frac{2a^3}{3} \int_0^{2\pi} \sin^2 \theta\, d\theta$$

$$= \frac{2a^3}{3} \int_0^{2\pi} \frac{1 - \cos 2\theta}{2}\, d\theta = \frac{2a^3}{3} \left[\frac{\theta}{2} - \frac{\sin 2\theta}{4} \right] \Bigg|_0^{2\pi} = \frac{2\pi a^3}{3}.$$

23. We have $\mathbf{T}_s = (\cos t, \sin t, 0)$ and $\mathbf{T}_t = (-s\sin t, s\cos t, 1)$, so that the standard normal is

$$\mathbf{N} = \mathbf{T}_s \times \mathbf{T}_t = \begin{vmatrix} \mathbf{i} & \mathbf{j} & \mathbf{k} \\ \cos t & \sin t & 0 \\ -s\sin t & s\cos t & 1 \end{vmatrix} = \sin t\, \mathbf{i} - \cos t\, \mathbf{j} + s\, \mathbf{k}.$$

Therefore, the flux of $\mathbf{F}$ is given by

$$\iint_S \mathbf{F} \cdot d\mathbf{S} = \int_0^{2\pi} \int_0^2 \mathbf{F}(\mathbf{X}(s,t)) \cdot \mathbf{N}(s,t)\, ds\, dt$$

$$= \int_0^{2\pi} \int_0^2 \left(s\sin t, s\cos t, t^3 \right) \cdot (\sin t, -\cos t, s)\, ds\, dt$$

$$= \int_0^{2\pi} \int_0^2 \left(s(\sin^2 t - \cos^2 t) + st^3 \right) ds\, dt$$

$$= \int_0^{2\pi} \int_0^2 \left(st^3 - s\cos 2t \right) ds\, dt = \int_0^{2\pi} \left(\tfrac{1}{2}s^2 t^3 - \tfrac{1}{2}s^2 \cos 2t \right) \Big|_{s=0}^{2} dt$$

$$= \int_0^{2\pi} \left(2t^3 - 2\cos 2t \right) dt = \left(\tfrac{1}{2}t^4 - \sin 2t \right) \Big|_0^{2\pi} = 8\pi^4.$$

27. (a) Below left is just the portion of S for $0 \le z \le 2$ so that you can more clearly see the funnel shape. Below right is a sketch of S.

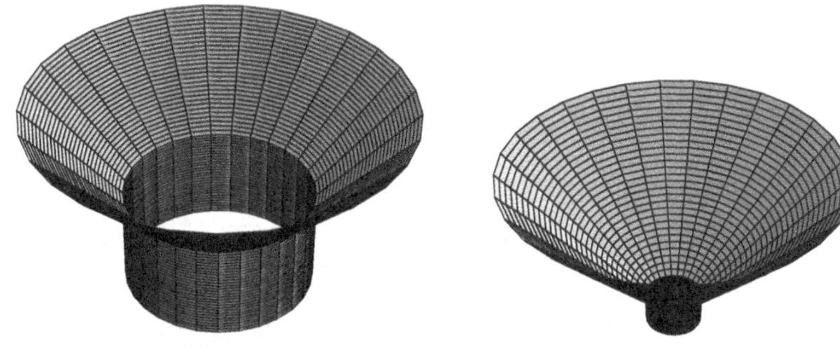

(b) For the cylindrical portion of S, $\mathbf{X}(s,t) = (\cos s, \sin s, t)$ for $0 \le s < 2\pi$ and $0 \le t \le 1$. In that case $\mathbf{T}_s(s,t) = (-\sin s, \cos s, 0)$, $\mathbf{T}_t(s,t) = (0,0,1)$, $\mathbf{T}_s(s,t) \times \mathbf{T}_t(s,t) = (\cos s, \sin s, 0)$ and so the outward pointing unit normal for this portion is $\mathbf{n} = (\cos s, \sin s, 0) = x\mathbf{i} + y\mathbf{j}$.
For the conical portion of S, $\mathbf{X}(s,t) = (t\cos s, t\sin s, t)$ for $0 \le s < 2\pi$ and $1 \le t \le 9$. In that case $\mathbf{T}_s(s,t) = (-t\sin s, t\cos s, 0)$, $\mathbf{T}_t(s,t) = (\cos s, \sin s, 1)$, $\mathbf{T}_s(s,t) \times \mathbf{T}_t(s,t) = (t\cos s, t\sin s, -t)$ and so the outward pointing unit normal for this portion is $\mathbf{n} = (1/\sqrt{2})(\cos s, \sin s, -1) = (1/\sqrt{2})((x/z)\mathbf{i} + (y/z)\mathbf{j} - \mathbf{k})$.

(c)

$$\iint_S \mathbf{F}\cdot d\mathbf{S} = \iint_S (-y\mathbf{i} + x\mathbf{j} + z\mathbf{k})\cdot d\mathbf{S} = \int_0^{2\pi}\int_1^9 (-t\sin s, t\cos s, t)\cdot(t\cos s, t\sin s, -t)\,dt\,ds$$

$$+\int_0^{2\pi}\int_0^1 (-\sin s, \cos s, t)\cdot(\cos s, \sin s, 0)\,dt\,ds = \int_0^{2\pi}\int_1^9 -t^2\,dt\,ds + \int_0^{2\pi}\int_0^1 0\,dt\,ds$$

$$= \int_0^{2\pi} -\frac{t^3}{3}\bigg|_1^9 ds = \int_0^{2\pi}\left[-\frac{729}{3} + \frac{1}{3}\right] ds = -\frac{1456\pi}{3}.$$

7.3 Stokes's and Gauss's Theorems

1. Calculate

$$\nabla\times\mathbf{F} = \begin{vmatrix} \mathbf{i} & \mathbf{j} & \mathbf{k} \\ \partial/\partial x & \partial/\partial y & \partial/\partial z \\ xz & yz & x^2+y^2 \end{vmatrix} = (2y-y)\mathbf{i} + (-2x+x)\mathbf{j} = y\mathbf{i} - x\mathbf{j}.$$

By symmetry we can see that the integral will be zero; however, let's follow the instructions. View the surface as a level set at height 1 of $f(x,y,z) = x^2+y^2+5z$ Then $\mathbf{N} = \nabla f = 2x\mathbf{i} + 2y\mathbf{j} + 5\mathbf{k}$. So,

$$\iint_S \nabla\times\mathbf{F}\cdot d\mathbf{S} = \iint_D (y\mathbf{i} - x\mathbf{j})\cdot(2x\mathbf{i} + 2y\mathbf{j} + 5\mathbf{k})\,dx\,dy$$

$$= \iint_D (2xy - 2xy)\,dx\,dy = 0.$$

On the other hand, ∂S consists of $C = \{(x,y,z)|x^2+y^2 = 1 \text{ and } z = 0\}$ which we parametrize by $\mathbf{x}(t) = (\cos t, \sin t, 0)$. Then,

$$\oint_{\partial S} \mathbf{F}\cdot d\mathbf{s} = \int_0^{2\pi} \mathbf{F}(\mathbf{x}(t))\cdot\mathbf{x}'(t)\,dt = \int_0^{2\pi} (0,0,1)\cdot(-\sin t, \cos t, 0)\,dt = 0.$$

These two answers agree.

5. Stokes's Theorem implies that we don't need to be concerned that S is defined as the union of S_1 and S_2 if we choose the calculation along the boundary. Then ∂S is parametrized by $\mathbf{x}(t) = (3\cos t, 3\sin t, 0)$ where $0 \le t \le 2\pi$, and so

$$\oint_{\partial S} \mathbf{F}\cdot d\mathbf{s} = \int_0^{2\pi} \mathbf{F}(\mathbf{x}(t))\cdot\mathbf{x}'(t)\,dt$$

$$= \int_0^{2\pi} (27\cos^3 t\mathbf{i} + 3^7\sin^7 t\mathbf{j})\cdot(-3\sin t, 3\cos t, 0)\,dt$$

$$= -3^4\int_0^{2\pi} \cos^3 t\sin t\,dt + 3^8\int_0^{2\pi} \sin^7 t\cos t\,dt = 0.$$

9. Since $\dfrac{\partial}{\partial x}\left(\dfrac{x}{\sqrt{x^2+y^2+z^2}}\right) = \dfrac{y^2+z^2}{(x^2+y^2+z^2)^{3/2}}$ we see that

$$\nabla\cdot\mathbf{F} = \frac{2(x^2+y^2+z^2)}{(x^2+y^2+z^2)^{3/2}} = \frac{2}{\sqrt{x^2+y^2+z^2}}, \quad \text{and}$$

$$\iiint_D \nabla\cdot\mathbf{F}\,dV = \iiint_D \frac{2}{\sqrt{x^2+y^2+z^2}}\,dV = \int_0^{2\pi}\int_0^{\pi}\int_a^b \frac{2}{\rho}(\rho^2\sin\varphi)\,d\rho\,d\varphi\,d\theta$$

$$= \int_0^{2\pi}\int_0^{\pi}\int_a^b (2\rho\sin\varphi)\,d\rho\,d\varphi\,d\theta = \int_0^{2\pi}\int_0^{\pi} [\rho^2\sin\varphi]\bigg|_a^b d\varphi\,d\theta$$

$$= (b^2-a^2)\int_0^{2\pi}\int_0^{\pi} \sin\varphi\,d\varphi\,d\theta = (b^2-a^2)\int_0^{2\pi} 2\,d\theta = 4\pi(b^2-a^2).$$

On the other hand the boundary consists of two pieces: S_1 is the sphere of radius a and S_2 is the sphere of radius b. Parametrize S_1 by $\mathbf{X}_1(s,t) = (a\sin s\cos t, a\sin s\sin t, a\cos s)$ for $0 \le s \le \pi$ and $0 \le t \le 2\pi$. Then a normal vector is $\mathbf{N}_1(s,t) = -a^2\sin s(\sin s\cos t, \sin s\sin t, \cos s)$. A similar calculation for S_2 yields $\mathbf{N}_2(s,t) = b^2\sin s(\sin s\cos t, \sin s\sin t, \cos s)$. Note that $\mathbf{N}_1$ is oriented pointing inward and $\mathbf{N}_2$ is oriented pointing outward. Then,

$$\oiint_S \mathbf{F}\cdot d\mathbf{S} = \iint_{S_1} \mathbf{F}\cdot d\mathbf{S} + \iint_{S_2} \mathbf{F}\cdot d\mathbf{S}$$

$$= \int_0^{2\pi}\int_0^{\pi} \frac{1}{a}((a\sin s\cos t, a\sin s\sin t, a\cos s)\cdot[-a^2\sin s(\sin s\cos t, \sin s\sin t, \cos s)]\,ds\,dt$$

$$+ \int_0^{2\pi}\int_0^{\pi} \frac{1}{b}(b\sin s\cos t, b\sin s\sin t, b\cos s)\cdot[b^2\sin s(\sin s\cos t, \sin s\sin t, \cos s)]\,ds\,dt$$

$$= \int_0^{2\pi}\int_0^{\pi}[-a^2\sin s]\,ds\,dt + \int_0^{2\pi}\int_0^{\pi}[b^2\sin s]\,ds\,dt$$

$$= \int_0^{2\pi}[-2a^2]\,dt + \int_0^{2\pi}[2b^2]\,dt = 4\pi(b^2 - a^2).$$

These two answers agree.

13. (a) By the double angle formula we have $z = \sin 2t = 2\sin t\cos t = 2xy$.

(b) $\oint_C (y^3 + \cos x)\,dx + (\sin y + z^2)\,dy + x\,dz = \oint_C \mathbf{F}\cdot d\mathbf{s}$ where $\mathbf{F} = (y^3 + \cos x)\mathbf{i} + (\sin y + z^2)\mathbf{j} + x\mathbf{k}$. By Stokes's theorem we may calculate the line integral by evaluating $\iint_S \nabla\times\mathbf{F}\cdot d\mathbf{S}$ where S is the portion of $z = 2xy$ bounded by C. (Note that S lies over the unit disk in the xy-plane.) Now $\nabla\times\mathbf{F} = -4xy\mathbf{i} - \mathbf{j} - 3y^2\mathbf{k} = -2z\mathbf{i} - \mathbf{j} - 3y^2\mathbf{k}$ on S. Note that the orientation of C is compatible with an upward orientation of S. So we may take for normal

$$\mathbf{n} = \frac{-2y\mathbf{i} - 2x\mathbf{j} + \mathbf{k}}{\sqrt{4x^2 + 4y^2 + 1}} \qquad \text{(unit normal of } S\text{)}.$$

Hence $\iint_S \nabla\times\mathbf{F}\cdot d\mathbf{S} = \iint_D (8xy^2 + 2x - 3y^2)\,dx\,dy$ (D = unit disk in xy-plane).
Now use polar coordinates, so that the integral becomes

$$\int_0^{2\pi}\int_0^1 (8r^3\sin^2\theta\cos\theta + 2r\cos\theta - 3r^2\sin^2\theta)r\,dr\,d\theta$$

$$= \int_0^{2\pi}\left(\frac{8}{5}\sin^2\theta\cos\theta + \frac{2}{3}\cos\theta - \frac{3}{4}\left(\frac{1}{2}(1-\cos 2\theta)\right)\right)d\theta$$

$$= \left(\frac{8}{15}\sin^3\theta + \frac{2}{3}\sin\theta - \frac{3}{8}\theta + \frac{3}{16}\sin 2\theta\right)\Bigg|_0^{2\pi} = -\frac{3\pi}{4}.$$

19. Let $\mathbf{X}\colon D \to \mathbf{R}^3$, $\mathbf{X}(u,v) = (x(u,v), y(u,v), z(u,v))$ parametrize S and $(u(t), v(t))$, $a \le t \le b$ parametrize ∂D so that $\mathbf{X}(u(t), v(t))$ parametrizes ∂S. (Note the assumption that ∂D can be parametrized by a single path—this is not a problem.) Write $\mathbf{F}$ as $M\mathbf{i} + N\mathbf{j} + P\mathbf{k}$. We need to show that

$$(*) \qquad \oint_{\partial S}(M\mathbf{i} + N\mathbf{j} + P\mathbf{k})\cdot d\mathbf{s} = \iint_S \left(\frac{\partial P}{\partial y} - \frac{\partial N}{\partial z}, \frac{\partial M}{\partial z} - \frac{\partial P}{\partial x}, \frac{\partial N}{\partial x} - \frac{\partial M}{\partial y}\right)\cdot d\mathbf{S}.$$

Consider the line integral in $(*)$. We may write it in differential form as $\oint_{\partial S} M\,dx + N\,dy + P\,dz$. Consider, for the moment, just the piece $\oint_{\partial S} M\,dx$. By the chain rule, $\dfrac{dx}{dt} = \dfrac{\partial x}{\partial u}\dfrac{du}{dt} + \dfrac{\partial x}{\partial v}\dfrac{dv}{dt}$. Hence,

$$\oint_{\partial S} M\,dx = \int_a^b M(\mathbf{X}(u(t), v(t)))\left(\frac{\partial x}{\partial u}\frac{du}{dt} + \frac{\partial x}{\partial v}\frac{dv}{dt}\right)dt = \int_{\partial D} M\circ\mathbf{X}\frac{\partial x}{\partial u}\,du + M\circ\mathbf{X}\frac{\partial x}{\partial v}\,dv.$$

The last line integral in just an integral in the uv-plane and so we may apply Green's theorem to find

$$\oint_{\partial S} M\,dx = \iint_D \left[\frac{\partial}{\partial u}\left(M\circ\mathbf{X}\frac{\partial x}{\partial v}\right) - \frac{\partial}{\partial v}\left(M\circ\mathbf{X}\frac{\partial x}{\partial u}\right)\right]du\,dv.$$

We need to apply the chain rule again, along with the product rule:

$$\frac{\partial}{\partial u}\left(M \circ \mathbf{X}\frac{\partial x}{\partial v}\right) = \left(\frac{\partial M}{\partial x}\frac{\partial x}{\partial u} + \frac{\partial M}{\partial y}\frac{\partial y}{\partial u} + \frac{\partial M}{\partial z}\frac{\partial z}{\partial u}\right)\frac{\partial x}{\partial v} + M \circ \mathbf{X}\frac{\partial^2 x}{\partial u\,\partial v}$$

$$\frac{\partial}{\partial v}\left(M \circ \mathbf{X}\frac{\partial x}{\partial u}\right) = \left(\frac{\partial M}{\partial x}\frac{\partial x}{\partial v} + \frac{\partial M}{\partial y}\frac{\partial y}{\partial v} + \frac{\partial M}{\partial z}\frac{\partial z}{\partial v}\right)\frac{\partial x}{\partial u} + M \circ \mathbf{X}\frac{\partial^2 x}{\partial v\,\partial u}.$$

Since the exercise allows us to assume that $\mathbf{X}$ is of class C^2, the mixed partials are equal: $\dfrac{\partial^2 x}{\partial u \partial v} = \dfrac{\partial^2 x}{\partial v \partial u}$. Therefore, our double integral becomes, after cancellation,

$$(**) \quad \iint_D \left[\frac{\partial M}{\partial y}\left(\frac{\partial x}{\partial v}\frac{\partial y}{\partial u} - \frac{\partial x}{\partial u}\frac{\partial y}{\partial v}\right) + \frac{\partial M}{\partial z}\left(\frac{\partial x}{\partial v}\frac{\partial z}{\partial u} - \frac{\partial x}{\partial u}\frac{\partial z}{\partial v}\right)\right] du\, dv.$$

Now consider the surface integral in $(*)$. Using the parametrization $\mathbf{X}$, and calculating the normal, we have that it is equal to

$$\iint_D (P_y - N_z, M_z - P_x, N_x - M_y) \cdot (y_u z_v - y_v z_u, z_u x_v - z_v x_u, x_u y_v - x_v y_u)\, du\, dv.$$

Next, calculate the dot product and isolate just those terms that contain M. Then the piece of the surface integral in $(*)$ that involves just M is

$$\iint_D \left[\frac{\partial M}{\partial z}\left(\frac{\partial z}{\partial u}\frac{\partial x}{\partial v} - \frac{\partial z}{\partial v}\frac{\partial x}{\partial u}\right) - \frac{\partial M}{\partial y}\left(\frac{\partial x}{\partial u}\frac{\partial y}{\partial v} - \frac{\partial x}{\partial v}\frac{\partial y}{\partial u}\right)\right] du\, dv.$$

This is the same as the double integral in $(**)$.

In an entirely analogous way, we may show that $\oint_{\partial S} N\,dy$ and $\oint_{\partial S} P dz$ are equal to the remaining pieces of the surface integral in $(*)$, completing the proof.

21. **(a)** If $\mathbf{F} = f\mathbf{a}$, then

$$\begin{aligned}\nabla \cdot \mathbf{F} &= \frac{\partial}{\partial x}(fa_1) + \frac{\partial}{\partial y}(fa_2) + \frac{\partial}{\partial z}(fa_3) \\ &= a_1\frac{\partial f}{\partial x} + a_2\frac{\partial f}{\partial y} + a_3\frac{\partial f}{\partial z} \\ &= \left(\frac{\partial f}{\partial x}, \frac{\partial f}{\partial y}, \frac{\partial f}{\partial z}\right) \cdot (a_1, a_2, a_3) = \nabla f \cdot \mathbf{a}.\end{aligned}$$

(b) With $\mathbf{F} = f\,\mathbf{i}$, we may apply Gauss's theorem:

$$\oiint_S \mathbf{F} \cdot d\mathbf{S} = \iiint_D \nabla \cdot \mathbf{F}\, dV.$$

The left side is

$$\oiint_S \mathbf{F} \cdot d\mathbf{S} = \oiint_S (\mathbf{F} \cdot \mathbf{n})\, dS = \oiint_S ((f\,\mathbf{i}) \cdot \mathbf{n})\, dS = \oiint_S (fn_1)\, dS.$$

Using part (a) with $\mathbf{a} = \mathbf{i}$, the right side is

$$\iiint_D \nabla \cdot \mathbf{F}\, dV = \iiint_D \nabla f \cdot \mathbf{i}\, dV = \iiint_D \frac{\partial f}{\partial x}\, dV.$$

Similarly, with $\mathbf{a} = \mathbf{j}$, we have

$$\oiint_S \mathbf{F} \cdot d\mathbf{S} = \oiint_S ((f\,\mathbf{j}) \cdot \mathbf{n})\, dS = \oiint_S (fn_2)\, dS$$

and

$$\iiint_D \nabla \cdot \mathbf{F}\, dV = \iiint_D \nabla f \cdot \mathbf{j}\, dV = \iiint_D \frac{\partial f}{\partial y}\, dV.$$

Finally, with $\mathbf{a} = \mathbf{k}$ we obtain

$$\oiint_S (fn_3)\, dS = \iiint_D \frac{\partial f}{\partial z}\, dV.$$

(c) Using part (b),

$$\oiint_S f\mathbf{n}\,dS = \left(\oiint_S fn_1\,dS, \oiint_S fn_2\,dS, \oiint_S fn_3\,dS\right)$$
$$= \left(\iiint_D \frac{\partial f}{\partial x}\,dV, \iiint_D \frac{\partial f}{\partial y}\,dV, \iiint_D \frac{\partial f}{\partial z}\,dV\right) = \iiint_D \nabla f\,dV.$$

25. (a) If $f(x,y,z) = \ln(x^2+y^2+z^2)$ then $\nabla f(x,y,z) = \dfrac{2x\mathbf{i}+2y\mathbf{j}+2z\mathbf{k}}{x^2+y^2+z^2} = \dfrac{2x\mathbf{i}+2y\mathbf{j}+2z\mathbf{k}}{a^2}$ on S. Also, the unit normal to the sphere that points away from the origin is $\mathbf{n}(x,y,z) = \dfrac{2x\mathbf{i}+2y\mathbf{j}+2z\mathbf{k}}{\sqrt{4(x^2+y^2+z^2)}} = \dfrac{2x\mathbf{i}+2y\mathbf{j}+2z\mathbf{k}}{2a} = \dfrac{x\mathbf{i}+y\mathbf{j}+z\mathbf{k}}{a}$. So,

$$\iint_S \frac{\partial f}{\partial n}\,dS = \iint_S \nabla f\cdot\mathbf{n}\,dS$$
$$= \iint_S \frac{2x\mathbf{i}+2y\mathbf{j}+2z\mathbf{k}}{x^2+y^2+z^2}\cdot\frac{x\mathbf{i}+y\mathbf{j}+z\mathbf{k}}{a}\,dS$$
$$= \iint_S \frac{2}{a}\,dS = \frac{2}{a}(\text{surface area of } S)$$
$$= \frac{2}{a}\left(\frac{4\pi a^2}{8}\right) = \pi a.$$

(b) First calculate that $\nabla\cdot(\nabla f) = \dfrac{2}{x^2+y^2+z^2}$. We'll use spherical coordinates to integrate.

$$\iiint_D \nabla\cdot(\nabla f)\,dV = \iiint_D \frac{2}{x^2+y^2+z^2}\,dV = \int_0^{\pi/2}\int_0^{\pi/2}\int_0^a \left(\frac{2}{\rho^2}\right)\rho^2\sin\varphi\,d\rho\,d\varphi\,d\theta$$
$$= 2\int_0^{\pi/2}\int_0^{\pi/2}\int_0^a \sin\varphi\,d\rho\,d\varphi\,d\theta = 2a\int_0^{\pi/2}\int_0^{\pi/2}\sin\varphi\,d\varphi\,d\theta$$
$$= 2a\int_0^{\pi/2} d\theta = \pi a.$$

(c) By Gauss's theorem, $\iiint_D \nabla\cdot(\nabla f)\,dV = \oiint_{\partial D}(\nabla f)\cdot d\mathbf{S}$. The boundary of D consists of four pieces: S, the surface from part (a); S_x, the intersection of D and the plane $x=0$; S_y, the intersection of D and the plane $y=0$; and S_z, the intersection of D and the plane $z=0$. On S_x we know that $\nabla f(0,y,z) = \dfrac{2y\mathbf{j}+2z\mathbf{k}}{y^2+z^2}$ and $\mathbf{n}=(-1,0,0)$ so

$$\iint_{S_x}\nabla f\cdot d\mathbf{S} = \iint_{S_x}\nabla f\cdot\mathbf{n}\,dS = \iint_{S_x} 0\,dS = 0.$$

A similar analysis gives us $\iint_{S_y}\nabla f\cdot d\mathbf{S}=0$ and $\iint_{S_z}\nabla f\cdot d\mathbf{S}=0$. Therefore,

$$\iiint_D \nabla\cdot(\nabla f)\,dV = \oiint_{\partial D}(\nabla f)\cdot d\mathbf{S} = \iint_S \nabla f\cdot d\mathbf{S} = \iint_S \frac{\partial f}{\partial n}\,dS.$$

29. (a) $\mathbf{F} = F_r\mathbf{e}_r + F_\theta\mathbf{e}_\theta + F_z\mathbf{e}_z$. The area of the top face is

$$(\Delta\theta/2\pi)[\pi(r+\Delta r/2)^2 - \pi(r-\Delta r/2)^2] = (\Delta\theta/2)(2r\Delta r) = r\Delta\theta\Delta r.$$

Therefore,

$$\iint_{\text{top}}\mathbf{F}\cdot d\mathbf{S} = \iint_{\text{top}}\mathbf{F}\cdot\mathbf{n}\,dS = \iint_{\text{top}}\mathbf{F}\cdot\mathbf{e}_z\,dS = \iint_{\text{top}} F_z\,dS$$
$$\approx F_z(r,\theta,z+\Delta z/2)(\text{area of top}) = F_z(r,\theta,z+\Delta z/2)r\Delta\theta\Delta r.$$

The calculation for the bottom face is similar. The differences are that the normal vector points down and F_z is evaluated at a different point. The result is that

$$\iint_{\text{bottom}} \mathbf{F} \cdot d\mathbf{S} \approx -F_z(r, \theta, z - \Delta z/2) r \Delta\theta \Delta r.$$

The area of the outer face is

$$(\Delta z)(\Delta\theta/2\pi)[2\pi(r + \Delta r/2)] = \Delta\theta\Delta z(r + \Delta r/2).$$

Therefore,

$$\begin{aligned}\iint_{\text{outer}} \mathbf{F} \cdot d\mathbf{S} &= \iint_{\text{outer}} \mathbf{F} \cdot \mathbf{n}\, dS = \iint_{\text{outer}} \mathbf{F} \cdot \mathbf{e}_r\, dS = \iint_{\text{outer}} F_r\, dS \\ &\approx F_r(r + \Delta r/2, \theta, z)(\text{area of outer}) = F_r(r + \Delta r/2, \theta, z)(r + \Delta r/2)\Delta\theta\Delta z.\end{aligned}$$

The calculation for the inner face is similar. The differences are that the normal vector points inward, F_r is evaluated at a different point, and the area of the face is slightly different. The result is that

$$\iint_{\text{inner}} \mathbf{F} \cdot d\mathbf{S} \approx -F_r(r - \Delta r/2, \theta, z)(r - \Delta r/2)\Delta\theta\Delta z.$$

The area of either the left or right face is just $\Delta r \Delta z$. Therefore, the integral along the left face (looking from the origin out at the solid) is

$$\begin{aligned}\iint_{\text{left}} \mathbf{F} \cdot d\mathbf{S} &= \iint_{\text{left}} \mathbf{F} \cdot \mathbf{n}\, dS = \iint_{\text{left}} \mathbf{F} \cdot \mathbf{e}_\theta\, dS = \iint_{\text{left}} F_\theta\, dS \\ &\approx F_\theta(r, \theta + \Delta\theta/2, z)(\text{area of left}) = F_\theta(r, \theta + \Delta\theta/2, z)\Delta r\Delta z.\end{aligned}$$

The calculation for the right face is similar. The differences are that the normal vector points the opposite direction and F_θ is evaluated at a different point. The result is that

$$\iint_{\text{right}} \mathbf{F} \cdot d\mathbf{S} \approx -F_\theta(r, \theta - \Delta\theta/2, z)\Delta r\Delta z.$$

We sum these to obtain

$$\begin{aligned}\oiint_S \mathbf{F} \cdot d\mathbf{S} \approx\ & F_z(r, \theta, z + \Delta z/2) r\Delta\theta\Delta r - F_z(r, \theta, z - \Delta z/2) r\Delta\theta\Delta r \\ &+ F_r(r + \Delta r/2, \theta, z)(r + \Delta r/2)\Delta\theta\Delta z - F_r(r - \Delta r/2, \theta, z)(r - \Delta r/2)\Delta\theta\Delta z \\ &+ F_\theta(r, \theta + \Delta\theta/2, z)\Delta r\Delta z - F_\theta(r, \theta - \Delta\theta/2, z)\Delta r\Delta z.\end{aligned}$$

(b) To calculate the divergence using the results of Exercise 27 we will divide the answer to part (a) by $V \approx r\Delta\theta\Delta r\Delta z$ and take the limit as $V \to 0$. Two notes before the calculation: 1) We can replace $\approx$ with $=$ because in the limit our approximation assumptions are true and 2) in evaluating each of the limits we use the remark given in the text at the end of Exercise 28 (although you may want to break the argument of the middle limit down further to see what is going on).

$$\begin{aligned}
\operatorname{div}\mathbf{F}(P) &= \lim_{V\to 0}\frac{1}{V}\oiint_S \mathbf{F}\cdot d\mathbf{S}\\
&= \lim_{V\to 0}\left[\frac{F_z(r,\theta,z+\Delta z/2)r\Delta\theta\Delta r - F_z(r,\theta,z-\Delta z/2)r\Delta\theta\Delta r}{r\Delta\theta\Delta r\Delta z}\right]\\
&\quad + \lim_{V\to 0}\left[\frac{F_r(r+\Delta r/2,\theta,z)(r+\Delta r/2)\Delta\theta\Delta z - F_r(r-\Delta r/2,\theta,z)(r-\Delta r/2)\Delta\theta\Delta z}{r\Delta\theta\Delta r\Delta z}\right]\\
&\quad + \lim_{V\to 0}\left[\frac{F_\theta(r,\theta+\Delta\theta/2,z)\Delta r\Delta z - F_\theta(r,\theta-\Delta\theta/2,z)\Delta r\Delta z}{r\Delta\theta\Delta r\Delta z}\right]\\
&= \lim_{\Delta z\to 0}\left[\frac{F_z(r,\theta,z+\Delta z/2) - F_z(r,\theta,z-\Delta z/2)}{\Delta z}\right]\\
&\quad + \lim_{\Delta r\to 0}\left[\frac{F_r(r+\Delta r/2,\theta,z)(r+\Delta r/2) - F_r(r-\Delta r/2,\theta,z)(r-\Delta r/2)}{r\Delta r}\right]\\
&\quad + \lim_{\Delta\theta\to 0}\left[\frac{F_\theta(r,\theta+\Delta\theta/2,z) - F_\theta(r,\theta-\Delta\theta/2,z)}{r\Delta\theta}\right]\\
&= \left[\frac{\partial F_z}{\partial z} + \frac{1}{r}\frac{\partial}{\partial r}(rF_r) + \frac{1}{r}\frac{\partial F_\theta}{\partial\theta}\right]\bigg|_P.
\end{aligned}$$

33. This is similar to Exercise 32. By Exercise 31, $\mathbf{e}_\rho\cdot\operatorname{curl}\mathbf{F}(P) = \lim_{A\to 0}\frac{1}{A}\oint_{C_A}\mathbf{F}\cdot d\mathbf{s}$. Here $A\approx \rho^2\sin\varphi\Delta\varphi\Delta\theta$.

$$\begin{aligned}
\oint_{C_A}\mathbf{F}\cdot d\mathbf{s} &= -F_\varphi\left(\rho,\theta+\frac{\Delta\theta}{2},\varphi\right)\rho\Delta\varphi - F_\theta\left(\rho,\theta,\varphi-\frac{\Delta\varphi}{2}\right)\rho\sin\left(\varphi-\frac{\Delta\varphi}{2}\right)\Delta\theta\\
&\quad + F_\varphi\left(\rho,\theta-\frac{\Delta\theta}{2},\varphi\right)\rho\Delta\varphi + F_\theta\left(\rho,\theta,\varphi+\frac{\Delta\varphi}{2}\right)\rho\sin\left(\varphi+\frac{\Delta\varphi}{2}\right)\Delta\theta.
\end{aligned}$$

Therefore,

$$\begin{aligned}
\mathbf{e}_\rho\cdot\operatorname{curl}\mathbf{F}(P) &= \lim_{A\to 0}\frac{1}{A}\oint_{C_A}\mathbf{F}\cdot d\mathbf{s}\\
&= \lim_{\Delta\theta\to 0}\left[-\frac{F_\varphi\left(\rho,\theta+\frac{\Delta\theta}{2},\varphi\right) - F_\varphi\left(\rho,\theta-\frac{\Delta\theta}{2},\varphi\right)}{\rho\sin\varphi\Delta\theta}\right]\\
&\quad + \lim_{\Delta\varphi\to 0}\left[\frac{F_\theta\left(\rho,\theta,\varphi+\frac{\Delta\varphi}{2}\right)\sin\left(\varphi+\frac{\Delta\varphi}{2}\right) - F_\theta\left(\rho,\theta,\varphi-\frac{\Delta\varphi}{2}\right)\sin\left(\varphi-\frac{\Delta\varphi}{2}\right)}{\rho\sin\varphi\Delta\varphi}\right]\\
&= \frac{1}{\rho\sin\varphi}\left[-\frac{\partial F_\varphi}{\partial\theta} + \frac{\partial}{\partial\varphi}(\sin\varphi F_\theta)\right].
\end{aligned}$$

Again, by Exercise 31, $\mathbf{e}_\theta\cdot\operatorname{curl}\mathbf{F}(P) = \lim_{A\to 0}\frac{1}{A}\oint_{C_A}\mathbf{F}\cdot d\mathbf{s}$. Here $A\approx\rho\Delta\varphi\Delta\rho$.

$$\begin{aligned}
\oint_{C_A}\mathbf{F}\cdot d\mathbf{s} &= F_\varphi\left(\rho+\frac{\Delta\rho}{2},\theta,\varphi\right)\left(\rho+\frac{\Delta\rho}{2}\right)\Delta\varphi - F_\rho\left(\rho,\theta,\varphi+\frac{\Delta\varphi}{2}\right)\Delta\rho\\
&\quad - F_\varphi\left(\rho-\frac{\Delta\rho}{2},\theta,\varphi\right)\left(\rho-\frac{\Delta\rho}{2}\right)\Delta\varphi + F_\rho\left(\rho,\theta,\varphi-\frac{\Delta\varphi}{2}\right)\Delta\rho.
\end{aligned}$$

Therefore,

$$\begin{aligned}
\mathbf{e}_\theta \cdot \operatorname{curl} \mathbf{F}(P) &= \lim_{A\to 0} \frac{1}{A} \oint_{C_A} \mathbf{F} \cdot d\mathbf{s} \\
&= \lim_{\Delta\rho\to 0} \left[\frac{F_\varphi\left(\rho + \frac{\Delta\rho}{2}, \theta, \varphi\right)\left(\rho + \frac{\Delta\rho}{2}\right) - F_\varphi\left(\rho - \frac{\Delta\rho}{2}, \theta, \varphi\right)\left(\rho - \frac{\Delta\rho}{2}\right)}{\rho\Delta\rho} \right] \\
&\quad + \lim_{\Delta\varphi\to 0} \left[-\frac{F_\rho\left(\rho, \theta, \varphi + \frac{\Delta\varphi}{2}\right) - F_\rho\left(\rho, \theta, \varphi - \frac{\Delta\varphi}{2}\right)}{\rho\Delta\varphi} \right] \\
&= \frac{1}{\rho}\left[\frac{\partial}{\partial\rho}(\rho F_\varphi) - \frac{\partial F_\rho}{\partial\varphi} \right].
\end{aligned}$$

Again, by Exercise 31, $\mathbf{e}_\varphi \cdot \operatorname{curl} \mathbf{F}(P) = \lim_{A\to 0} \dfrac{1}{A} \oint_{C_A} \mathbf{F} \cdot d\mathbf{s}$. Here $A \approx \rho \sin\varphi \Delta\rho\Delta\theta$.

$$\begin{aligned}
\oint_{C_A} \mathbf{F} \cdot d\mathbf{s} &= F_\theta\left(\rho - \frac{\Delta\rho}{2}, \theta, \varphi\right)\left(\rho - \frac{\Delta\rho}{2}\right) \sin\varphi\Delta\theta + F_\rho\left(\rho, \theta + \frac{\Delta\theta}{2}, \varphi\right)\Delta\rho \\
&\quad - F_\theta\left(\rho + \frac{\Delta\rho}{2}, \theta, \varphi\right)\left(\rho + \frac{\Delta\rho}{2}\right) \sin\varphi\Delta\theta - F_\rho\left(\rho, \theta - \frac{\Delta\theta}{2}, \varphi\right)\Delta\rho.
\end{aligned}$$

Therefore,

$$\begin{aligned}
\mathbf{e}_\varphi \cdot \operatorname{curl} \mathbf{F}(P) &= \lim_{A\to 0} \frac{1}{A} \oint_{C_A} \mathbf{F} \cdot d\mathbf{s} \\
&= \lim_{\Delta\rho\to 0} \left[-\frac{-F_\theta\left(\rho + \frac{\Delta\rho}{2}, \theta, \varphi\right)\left(\rho + \frac{\Delta\rho}{2}\right) + F_\theta\left(\rho - \frac{\Delta\rho}{2}, \theta, \varphi\right)\left(\rho - \frac{\Delta\rho}{2}\right)}{\rho\Delta\rho} \right] \\
&\quad + \lim_{\Delta\theta\to 0} \left[\frac{F_\rho\left(\rho, \theta + \frac{\Delta\theta}{2}, \varphi\right) - F_\rho\left(\rho, \theta - \frac{\Delta\theta}{2}, \varphi\right)}{\rho \sin\varphi \Delta\theta} \right] \\
&= \frac{1}{\rho}\left[-\frac{\partial}{\partial\rho}(\rho F_\theta) + \frac{1}{\sin\varphi}\frac{\partial F_\rho}{\partial\theta} \right].
\end{aligned}$$

Again, the final conclusion is just a matter of assembling the pieces above and checking that the sum agrees with the determinant.

7.4 Further Vector Analysis; Maxwell's Equations

3. (a) Using Green's first formula with $f = g$, we obtain

$$\iiint_D \nabla f \cdot \nabla f \, dV + \iiint_D f \nabla^2 f \, dV = \oint\!\!\!\oint_S f \nabla f \cdot d\mathbf{S}.$$

We are assuming that f is harmonic, so the second integral on the left side is 0. Therefore,

$$\iiint_D \nabla f \cdot \nabla f \, dV = \oint\!\!\!\oint_{\partial D} f \nabla f \cdot d\mathbf{S} = \oint\!\!\!\oint_{\partial D} f(\nabla f \cdot \mathbf{n}) \, dS = \oint\!\!\!\oint_{\partial D} f \frac{\partial f}{\partial n} \, dS.$$

(b) If $f = 0$ on the boundary of D, then part (a) implies that

$$0 = \oint\!\!\!\oint_{\partial D} f \frac{\partial f}{\partial n} \, dS = \iiint_D \nabla f \cdot \nabla f \, dV.$$

But $\nabla f \cdot \nabla f = \|\nabla f\|^2 \geq 0$. So the right-hand integral was of a non-negative, continuous integrand. For this to be zero, the integrand must have been identically zero. In other words, $\nabla f \cdot \nabla f$ is zero on D. We conclude that ∇f is zero on D and so f is constant on D. Since $f(x, y, z) = 0$ on ∂D and f is constant on D, we must have that $f \equiv 0$ on D.

7. We know from the argument in Exercise 6 that $-\iiint_D \sigma\rho\frac{\partial T}{\partial t}dV = \iiint_D -\nabla\cdot(k\nabla T)dV$. Use the product rule to conclude that this equals $\iiint_D -(\nabla k\cdot\nabla T + k\nabla^2 T)dV$. As before, shrink to a point to conclude $\sigma\rho\frac{\partial T}{\partial t} = k\nabla^2 T + \nabla k\cdot\nabla T$.

11. From Ampère's law we have $\mathbf{J} = \frac{1}{\mu_0}\nabla\times\mathbf{B} - \epsilon_0\frac{\partial \mathbf{E}}{\partial t}$. Therefore,

$$\begin{aligned}\nabla\times\mathbf{J} &= \frac{1}{\mu_0}\nabla\cdot(\nabla\times\mathbf{B}) - \epsilon_0\nabla\cdot\frac{\partial\mathbf{E}}{\partial t} = -\epsilon_0\nabla\cdot\frac{\partial\mathbf{E}}{\partial t}\\ &= -\epsilon_0\frac{\partial}{\partial t}(\nabla\cdot\mathbf{E}) = -\epsilon_0\frac{\partial}{\partial t}\left(\frac{\rho}{\epsilon_0}\right) \quad \text{by Gauss's law,}\\ &= -\frac{\partial\rho}{\partial t}.\end{aligned}$$

15. **(a)** This is just a straightforward calculation. Write $\mathbf{F} = M\mathbf{i} + N\mathbf{j} + P\mathbf{k}$. Then

$$\nabla\times\mathbf{F} = \begin{vmatrix}\mathbf{i} & \mathbf{j} & \mathbf{k}\\ \partial/\partial x & \partial/\partial y & \partial/\partial z\\ M & N & P\end{vmatrix} = (P_y - N_z)\mathbf{i} + (M_z - P_x)\mathbf{j} + (N_x - M_y)\mathbf{k}$$

and

$$\begin{aligned}\nabla\times(\nabla\times\mathbf{F}) &= \begin{vmatrix}\mathbf{i} & \mathbf{j} & \mathbf{k}\\ \partial/\partial x & \partial/\partial y & \partial/\partial z\\ P_y - N_z & M_z - P_x & N_x - M_y\end{vmatrix}\\ &= (N_{xy} - M_{yy} - M_{zz} + P_{xz})\mathbf{i} + (P_{yz} - N_{zz} - N_{xx} + M_{yx})\mathbf{j}\\ &\quad + (M_{zx} - P_{xx} - P_{yy} + N_{zy})\mathbf{k}.\end{aligned}$$

On the other hand,

$$\begin{aligned}\nabla(\nabla\cdot\mathbf{F}) &= \nabla(M_x + N_y + P_z)\\ &= (M_{xx} + N_{yx} + P_{zx})\mathbf{i} + (M_{xy} + N_{yy} + P_{zy})\mathbf{j} + (M_{xz} + N_{yz} + P_{zz})\mathbf{k}\\ \text{and } (\nabla\cdot\nabla)\mathbf{F} &= \left(\frac{\partial^2}{\partial x^2} + \frac{\partial^2}{\partial y^2} + \frac{\partial^2}{\partial z^2}\right)\mathbf{F}\\ &= (M_{xx} + M_{yy} + M_{zz})\mathbf{i} + (N_{xx} + N_{yy} + N_{zz})\mathbf{j} + (P_{xx} + P_{yy} + P_{zz})\mathbf{k}.\\ \text{Hence, } \nabla(\nabla\cdot\mathbf{F}) - \nabla^2\mathbf{F} &= (N_{yx} + P_{zx} - M_{yy} - M_{zz})\mathbf{i} + (M_{xy} + P_{zy} - N_{xx} - N_{zz})\mathbf{j}\\ &\quad + (M_{xz} + N_{yz} - P_{xx} - P_{yy})\mathbf{k}.\end{aligned}$$

By assumption $\mathbf{F}$ is of class C^2 and so the mixed partials are equal; thus we have the result:

$$\nabla\times(\nabla\times\mathbf{F}) = \nabla(\nabla\cdot\mathbf{F}) - \nabla^2\mathbf{F}.$$

(b) First we show that $\mathbf{E}$ satisfies the wave equation.

$$\begin{aligned}\nabla^2\mathbf{E} &= \nabla(\nabla\cdot\mathbf{E}) - \nabla\times(\nabla\times\mathbf{E}) \quad \text{from part (a),}\\ &= \nabla\left(\frac{\rho}{\epsilon_0}\right) - \nabla\times\left(-\frac{\partial\mathbf{B}}{\partial t}\right) \quad \text{using Gauss's and Faraday's laws,}\\ &= \frac{1}{\epsilon_0}\nabla\rho + \frac{\partial}{\partial t}(\nabla\times\mathbf{B})\\ &= \frac{1}{\epsilon_0}\nabla\rho + \frac{\partial}{\partial t}\left(\mu_0\mathbf{J} + \epsilon_0\,\mu_0\frac{\partial\mathbf{E}}{\partial t}\right) \quad \text{using Ampère's law}\\ &= \mathbf{0} + \epsilon_0\mu_0\frac{\partial^2\mathbf{E}}{\partial t^2} \quad \text{since there are no charges or currents (so } \rho\equiv 0 \text{ and } \mathbf{J}\equiv\mathbf{0}).\end{aligned}$$

Thus $\nabla^2 \mathbf{E} = k \dfrac{\partial^2 \mathbf{E}}{\partial t^2}$ where $k = \epsilon_0 \mu_0$.

Next we show that $\mathbf{B}$ satisfies the wave equation.

$$\begin{aligned}
\nabla^2 \mathbf{B} &= \nabla(\nabla \cdot \mathbf{B}) - \nabla \times (\nabla \times \mathbf{B}) \quad \text{from part (a),} \\
&= \mathbf{0} - \nabla \times \left(\mu_0 \mathbf{J} + \epsilon_0 \mu_0 \frac{\partial \mathbf{E}}{\partial t} \right) \quad \text{using Maxwell's equations,} \\
&= -\epsilon_0 \mu_0 \nabla \times \frac{\partial \mathbf{E}}{\partial t} \text{ since } \mathbf{J} \equiv \mathbf{0} \quad \text{(no currents),} \\
&= -\epsilon_0 \mu_0 \frac{\partial}{\partial t} (\nabla \times \mathbf{E}) \\
&= -\epsilon_0 \mu_0 \frac{\partial}{\partial t} \left(-\frac{\partial \mathbf{B}}{\partial t} \right) \quad \text{from Faraday's law,} \\
&= \epsilon_0 \mu_0 \frac{\partial^2 \mathbf{B}}{\partial t^2}.
\end{aligned}$$

So $\nabla^2 \mathbf{B} = k \dfrac{\partial^2 B}{\partial t^2}$ where $k = \epsilon_0 \mu_0$.

(c) By part (a),

$$\begin{aligned}
\nabla(\nabla \cdot \mathbf{E}) - (\nabla \cdot \nabla)\mathbf{E} &= \nabla \times (\nabla \times \mathbf{E}) \\
&= \nabla \times \left(-\frac{\partial \mathbf{B}}{\partial t} \right) \quad \text{by Faraday's law,} \\
&= -\frac{\partial}{\partial t} (\nabla \times \mathbf{B}) \\
&= -\frac{\partial}{\partial t} \left[\mu_0 \mathbf{J} + \epsilon_0 \mu_0 \frac{\partial \mathbf{E}}{\partial t} \right] \quad \text{by Ampère's law,} \\
&= -\mu_0 \frac{\partial}{\partial t} \left[\mathbf{J} + \epsilon_0 \frac{\partial \mathbf{E}}{\partial t} \right].
\end{aligned}$$

(d) Again from part (a),

$$\begin{aligned}
\nabla^2 \mathbf{E} &= \nabla(\nabla \cdot \mathbf{E}) - \nabla \times (\nabla \times \mathbf{E}) \\
&= \mathbf{0} - \nabla \times (\nabla \times \mathbf{E}) \quad \text{by Gauss's law and the fact that } \rho = 0, \\
&= \mu_0 \frac{\partial \mathbf{J}}{\partial t} + \epsilon_0 \mu_0 \frac{\partial^2 \mathbf{E}}{\partial t^2} \quad \text{from the argument in part (c).}
\end{aligned}$$

19. Since $\mathbf{P} = \mathbf{E} \times \mathbf{B}$,

$$\begin{aligned}
\oint\!\!\!\oint_S \mathbf{P} \cdot d\mathbf{S} &= \oint\!\!\!\oint_S (\mathbf{E} \times \mathbf{B}) \cdot d\mathbf{S} \\
&= \iiint_D \nabla \cdot (\mathbf{E} \times \mathbf{B}) \, dV \quad \text{by Gauss's theorem,} \\
&= \iiint_D (\mathbf{B} \cdot (\nabla \times \mathbf{E}) - \mathbf{E} \cdot (\nabla \times \mathbf{B})) \, dV \\
&= \iiint_D \left[-\mathbf{B} \cdot \frac{\partial \mathbf{B}}{\partial t} - \mathbf{E} \cdot \left(\mu_0 \mathbf{J} + \epsilon_0 \mu_0 \frac{\partial \mathbf{E}}{\partial t} \right) \right] dV \quad \text{from Faraday and Ampère's laws.}
\end{aligned}$$

Since $\mathbf{B}$ and $\mathbf{E}$ are both assumed to be constant in time, $\dfrac{\partial \mathbf{B}}{\partial t} = \dfrac{\partial \mathbf{E}}{\partial t} = \mathbf{0}$. Therefore, we get the desired result:

$$\oint\!\!\!\oint_S \mathbf{P} \cdot d\mathbf{S} = \iiint_D -\mu_0 \mathbf{E} \cdot \mathbf{J} \, dV.$$

True/False Exercises for Chapter 7

1. True.
3. True. (Let $u = s^3$ and $v = \tan t$.)
5. False. (The limits of integration are not correct.)
7. False. (The value of the integral is 24.)
9. True.
11. False. (The integral has value 32π.)
13. False. (The value is 0.)
15. False. (The surface must be connected.)
17. True. (The result follows from Stokes's theorem.)
19. True. (Use Gauss's theorem.)
21. False. (Gauss's theorem implies that the integral is *at most* twice the surface area.)
23. True.
25. False. (Should be the flux of the *curl* of $\mathbf{F}$.)
27. True. (Apply Green's first formula.)
29. False. (f is determined up to addition of a harmonic function.)

Miscellanenous Exercises for Chapter 7

3. (a) If we use cylindrical coordinates $x = r\cos\theta, y = r\sin\theta, z = z$, then the equation $x^2+y^2-z^2 = 1$ becomes $r^2-z^2 = 1$ or, since $r \geq 0, r = \sqrt{z^2+1}$. Hence the desired parametrization is $\mathbf{X}(z,\theta) = (\sqrt{z^2+1}\cos\theta, \sqrt{z^2+1}\sin\theta, z)$ where $z \in \mathbf{R}$ and $0 \leq \theta \leq 2\pi$.

(b) Modify the cylindrical coordinate substitution by letting $x = ar\cos t, y = br\sin t, z = cs$. Substitution into the equation for the hyperboloid yields $r^2 - s^2 = 1$ so $r = \sqrt{s^2+1}$. Hence a parametrization is $\mathbf{X}(s,t) = (a\sqrt{s^2+1}\cos t,$ $b\sqrt{s^2+1}\sin t, cs)$, where $s \in \mathbf{R}$ and $0 \leq t \leq 2\pi$.

(c) Substitute the parametric equations for $\mathbf{l}_1$ into the left side of the equation for the hyperboloid:

$$\begin{aligned}\frac{a^2(x_0-y_0t)^2}{a^2}+\frac{b^2(x_0t+y_0)^2}{b^2}-\frac{c^2t^2}{c^2} &= x_0^2-2x_0y_0t+y_0^2t^2+x_0^2t^2+2x_0y_0t+y_0^2-t^2\\ &= x_0^2+y_0^2+(x_0^2+y_0^2)t^2-t^2\\ &= 1+t^2-t^2 = 1,\end{aligned}$$

since $x_0^2+y_0^2 = 1$. Thus $\mathbf{l}_1$ lies on the hyperboloid. The calculation for $\mathbf{l}_2$ is similar.

(d) The plane tangent to the hyperboloid at the point $(ax_0, by_0, 0)$ is given by

$$\nabla F(ax_0, by_0, 0)\cdot(x-ax_0, y-by_0, 0) = 0 \quad \text{where} \quad F = \frac{x^2}{a^2}+\frac{y^2}{b^2}-\frac{z^2}{c^2}.$$

That is, the tangent plane is

$$(*) \quad \frac{x_0}{a}(x-ax_0)+\frac{y_0}{b}(y-by_0) = 0.$$

If we substitute the parametric equations for $\mathbf{l}_1$ into the left side of $(*)$, we find

$$\frac{x_0}{a}(a(x_0-y_0t)-ax_0)+\frac{y_0}{b}(b(x_0t+y_0)-by_0) = -x_0y_0t+y_0x_0t = 0$$

for all t. Therefore, the line $\mathbf{l}_1$ lies in the plane. A similar calculation can be made for $\mathbf{l}_2$.

7. (a) This should remind you of using washer and shell methods for surfaces of revolution. Of course, here we are finding a surface area and not volume. If you look at the specific value $s = s_0$ then, since we are revolving around the y-axis, we are sweeping out a circle of radius s_0 in the plane $y = f(s_0)$. Therefore, a parametrization is $\mathbf{X}(s,t) = (s\cos t, f(s), s\sin t), a \leq s \leq b, 0 \leq t \leq 2\pi$. Compare this with Exercise 6(a).

(b) Using the parametrization in (a), $\mathbf{T}_s = (\cos t, f'(s), \sin t)$ and $\mathbf{T}_t = (-s\sin t, 0, s\cos t)$. Therefore, $\mathbf{N} = \mathbf{T}_s \times \mathbf{T}_t = (sf'(s)\cos t, -s, sf'(s)\sin t)$, so $\|\mathbf{N}\| = \sqrt{s^2[f'(s)]^2 + s^2}$. Hence

$$\begin{aligned}\text{Surface area} &= \int_a^b \int_0^{2\pi} \sqrt{s^2[f'(s)]^2 + s^2}\, dt\, ds \\ &= 2\pi \int_a^b s\sqrt{[f'(s)]^2 + 1}\, ds \\ &= 2\pi \int_a^b x\sqrt{[f'(x)]^2 + 1}\, dx\end{aligned}$$

by changing the variable of integration. Compare this result with that of Exercise 6(b).

9. (a) The surface integral $\iint_S f dS$, roughly speaking, represents the "sum" of all the values of f on S. The area of S is a measure of the size of S. So the quotient can be thought of as the "total" amount of f divided by the size of the region being sampled.

(b) Parametrize the sphere as $\mathbf{X}(s,t) = (7\cos s\sin t, 7\sin s\sin t, 7\cos t)$, $0 \le s \le 2\pi$, $0 \le t \le \pi$. Then, following Example 11 in Section 7.1, $\|\mathbf{T}_s \times \mathbf{T}_t\| = 49\sin t$. Note that, on the surface S, the temperature $T(x,y,z) = x^2+y^2-3z = 49 - z^2 - 3z$. As a result, we can calculate

$$\begin{aligned}\iint_S T(x,y,z)\, dS &= \int_0^{2\pi} \int_0^{\pi} (49 - 49\cos^2 t - 21\cos t)49\sin t\, dt\, ds \\ &= 49\int_0^{2\pi} \left(-49\cos t + \frac{49}{3}\cos^3 t + \frac{21}{2}\cos^2 t\right)\Big|_0^{\pi} ds \\ &= 49\int_0^{2\pi} \left(49(2) - \frac{49}{3}(2) + \frac{21}{2}(1-1)\right) ds = \frac{(49)^2(4)(2\pi)}{3}.\end{aligned}$$

Now, since the surface area of a sphere of radius 7 is $4\pi(49)$, we have

$$[T]_{\text{avg}} = \frac{\iint_S T(x,y,z)\, dS}{\text{surface area}} = \frac{(49)^2(4)(2\pi)}{3}\frac{1}{4\pi(49)} = \frac{98}{3}.$$

13. By the symmetry of the surface, we must have $\bar{x} = \bar{y} = \bar{z}$. We compute $\bar{z}$ explicitly. Since δ is constant, it will cancel from the center of mass integrals:

$$\bar{z} = \frac{\iint_S z\delta\, dS}{\iint_S \delta\, dS} = \frac{\delta\iint_S z\, dS}{\delta\iint_S dS} = \frac{\iint_S z\, dS}{\text{surface area of } S}.$$

The surface area of the first octant portion of a sphere of radius a is $\frac{1}{8}(4\pi a^2) = \frac{1}{2}\pi a^2$. Therefore, $\bar{z} = \frac{2}{\pi a^2}\iint_S z\, dS$. We may parametrize the first octant portion of the sphere as $\mathbf{X}(\varphi,\theta) = (a\sin\varphi\cos\theta, a\sin\varphi\sin\theta, a\cos\varphi)$, $0 \le \varphi \le \pi/2$, $0 \le \theta \le \pi/2$. Hence,

$$\begin{aligned}\mathbf{T}_\varphi &= (a\cos\varphi\cos\theta, a\cos\varphi\sin\theta, -a\sin\varphi), \\ \mathbf{T}_\theta &= (-a\sin\varphi\sin\theta, a\sin\varphi\cos\theta, 0).\end{aligned}$$

Therefore,

$$\mathbf{N} = (a^2\sin^2\varphi\cos\theta, a^2\sin^2\varphi\sin\theta, a^2\sin\varphi\cos\theta) \quad \text{and} \quad \|\mathbf{N}\| = a^2\sin\varphi.$$

Thus,

$$\begin{aligned}\bar{z} &= \frac{2}{\pi a^2}\int_0^{\pi/2}\int_0^{\pi/2} (a\cos\varphi)a^2\sin\varphi\, d\varphi\, d\theta \\ &= \frac{2a^3}{\pi a^2}\int_0^{\pi/2}\int_0^{\pi/2} \cos\varphi\sin\varphi\, d\varphi\, d\theta \\ &= \frac{2a}{\pi}\int_0^{\pi/2} \left(\frac{1}{2}\sin^2\varphi\right)\Big|_{\varphi=0}^{\pi/2} d\theta \\ &= \frac{2a}{\pi}\int_0^{\pi/2} \frac{1}{2}\, d\theta = \frac{a}{\pi}\cdot\frac{\pi}{2} = \frac{a}{2}.\end{aligned}$$

17. (a) Parametrize the frustum $z^2 = 4x^2 + 4y^2$, $2 \le z \le 4$, as $\mathbf{X}(r, \theta) = (r\cos\theta, r\sin\theta, 2r)$, $0 \le \theta \le 2\pi$, $1 \le r \le 2$. Then

$$\frac{\partial(x,y)}{\partial(r,\theta)} = \begin{vmatrix} \cos\theta & -r\sin\theta \\ \sin\theta & r\cos\theta \end{vmatrix} = r$$

$$\frac{\partial(x,z)}{\partial(r,\theta)} = \begin{vmatrix} \cos\theta & -r\sin\theta \\ 2 & 0 \end{vmatrix} = 2r\sin\theta$$

$$\frac{\partial(y,z)}{\partial(r,\theta)} = \begin{vmatrix} \sin\theta & r\cos\theta \\ 2 & 0 \end{vmatrix} = -2r\cos\theta.$$

Therefore,

$$\begin{aligned} I_z &= \iint_S (x^2 + y^2)\, dS = \int_0^{2\pi} \int_1^2 r^2 \sqrt{r^2 + 4r^2}\, dr\, d\theta \\ &= \int_0^{2\pi} \int_1^2 \sqrt{5} r^3\, dr\, d\theta = \int_0^{2\pi} \frac{\sqrt{5}}{4}(15)\, d\theta = \frac{15\sqrt{5}\pi}{2}. \end{aligned}$$

(b) The radius of gyration is given by $r_z = \sqrt{\dfrac{I_z}{M}}$. Assuming, as in part (a), that the density is 1, the total mass is just the surface area of the frustum. This can be computed from the surface area of the cone without much trouble. We view the frustum as a large cone (of height 4) with the tip (a similar cone of height 2) removed and note that the surface area of a cone is π(radius)(slant height). Then

$$\text{Surface area of frustum} = \pi(2)(2\sqrt{5}) - \pi(1)(\sqrt{5}) = 3\sqrt{5}\pi.$$

Hence

$$r_z = \sqrt{\frac{I_z}{M}} = \sqrt{\frac{15\sqrt{5}\pi}{2}\frac{1}{3\sqrt{5}\pi}} = \sqrt{\frac{5}{2}}.$$

(Note: you can also compute the surface area as $\displaystyle\int_0^{2\pi}\int_1^2 \sqrt{5} r\, dr\, d\theta$.)

(c) We recompute the integral for I_z with $\delta = kr$. Thus

$$\begin{aligned} I_z &= \iint_S (x^2 + y^2)\delta\, dS \\ &= \int_0^{2\pi}\int_1^2 r^2 kr\sqrt{5r^2}\, dr\, d\theta \\ &= \int_0^{2\pi}\int_1^2 \sqrt{5} k r^4\, dr\, d\theta \\ &= \int_0^{2\pi} \frac{\sqrt{5}k}{5}(2^5 - 1)\, d\theta = \frac{62\sqrt{5}\pi k}{5}. \end{aligned}$$

The total mass of the frustum is

$$\begin{aligned} M &= \iint_S \delta\, dS = \int_0^{2\pi}\int_1^2 kr\sqrt{5} r\, dr\, d\theta \\ &= \int_0^{2\pi} \frac{\sqrt{5}}{3}(2^3 - 1)\, d\theta = \frac{14\sqrt{5}\pi k}{3}. \end{aligned}$$

Hence

$$r_z = \sqrt{\frac{I_z}{M}} = \sqrt{\frac{62\sqrt{5}\pi k}{5}\frac{3}{14\sqrt{5}\pi k}} = \sqrt{\frac{93}{35}}.$$

21. **(a)** Let $\mathbf{a} = (a_1, a_2, a_3)$ and assume $\mathbf{x}(t) = (x(t), y(t), z(t))$ parametrizes C. Then

$$\begin{aligned}\oint_C \mathbf{a} \cdot d\mathbf{s} &= \int_a^b \mathbf{a} \cdot \mathbf{x}'(t)\, dt \\ &= \int_a^b (a_1 x'(t) + a_2 y'(t) + a_3 z'(t))\, dt \\ &= (a_1 x(t) + a_2 y(t) + a_3 z(t))\Big|_a^b \\ &= \mathbf{a} \cdot \mathbf{x}(t)\Big|_a^b = \mathbf{a} \cdot \mathbf{x}(b) - \mathbf{a} \cdot \mathbf{x}(a) \\ &= 0\end{aligned}$$

since $\mathbf{x}(a) = \mathbf{x}(b)$ because C is a closed curve.

(b) Let S be any smooth, orientable surface with boundary curve C. If we orient S appropriately and use Stokes's theorem, we have

$$\oint_C \mathbf{a} \cdot d\mathbf{s} = \iint_S \nabla \times \mathbf{a} \cdot d\mathbf{S} = \iint_S \mathbf{0} \cdot d\mathbf{S} = 0.$$

25.

$$\begin{aligned}\oint_{\partial S} (f\nabla f) \cdot d\mathbf{s} &= \oint_{\partial S} \frac{1}{2}(f\nabla f + f\nabla f) \cdot d\mathbf{s} \\ &= 0 \quad \text{by Exercise 24.}\end{aligned}$$

29. Note that the boundary ∂W of W consists of three parts: S, $\tilde{S}_a$ and the lateral surfaces L of ∂W. With ∂W oriented by outward normal, and if we take S and $\tilde{S}_a$ to be oriented in the same way,

$$\oiint_{\partial W} \frac{\mathbf{x}}{\|\mathbf{x}\|^3} \cdot d\mathbf{S} = \pm\left(\iint_S \frac{\mathbf{x}}{\|\mathbf{x}\|^3} \cdot d\mathbf{S} - \iint_{\tilde{S}_a} \frac{\mathbf{x}}{\|\mathbf{x}\|^3} \cdot d\mathbf{S}\right) + \iint_L \frac{\mathbf{x}}{\|\mathbf{x}\|^3} \cdot d\mathbf{S}$$

(The $\pm$ sign depends on how S, $\tilde{S}_a$ are oriented with respect to the orientation of ∂W.) Now L consists of a collection of segments of the rays defining $\Omega(S, O)$. Thus L is *tangent* to $\mathbf{x}$. Hence $\mathbf{x} \cdot \mathbf{n} = 0$ where $\mathbf{n}$ is the appropriate unit normal to L. Thus $\iint_L \frac{\mathbf{x}}{\|\mathbf{x}\|^3} \cdot d\mathbf{S} = 0$. Thus

$$\oiint_{\partial W} \frac{\mathbf{x}}{\|\mathbf{x}\|^3} \cdot d\mathbf{S} = \pm\left(\iint_S \frac{\mathbf{x}}{\|\mathbf{x}\|^3} \cdot d\mathbf{S} - \iint_{\tilde{S}_a} \frac{\mathbf{x}}{\|\mathbf{x}\|^3} \cdot d\mathbf{S}\right).$$

Gauss's theorem implies

$$\oiint_{\partial W} \frac{\mathbf{x}}{\|\mathbf{x}\|^3} \cdot d\mathbf{S} = \iiint_W \left(\nabla \cdot \frac{\mathbf{x}}{\|\mathbf{x}\|^3}\right) dV = \iiint_W 0\, dV.$$

Hence $\iint_S \frac{\mathbf{x}}{\|\mathbf{x}\|^3} \cdot d\mathbf{S} = \iint_{\tilde{S}_a} \frac{\mathbf{x}}{\|\mathbf{x}\|^3} \cdot d\mathbf{S}$. On $\tilde{S}_a$, $\mathbf{n} = \frac{\mathbf{x}}{\|\mathbf{x}\|}$, so

$$\begin{aligned}\Omega(S, O) &= \iint_S \frac{\mathbf{x}}{\|\mathbf{x}\|^3} \cdot d\mathbf{S} = \iint_{\tilde{S}_a} \frac{\mathbf{x}}{\|\mathbf{x}\|^3} \cdot \frac{\mathbf{x}}{\|\mathbf{x}\|}\, dS \\ &= \iint_{\tilde{S}_a} \frac{\|\mathbf{x}\|^2}{\|\mathbf{x}\|^4}\, dS = \iint_{\tilde{S}_a} \frac{1}{\|\mathbf{x}\|^2}\, dS.\end{aligned}$$

But on $\tilde{S}_a$, $\|\mathbf{x}\| = a$, so

$$\Omega(S, O) = \iint_{\tilde{S}_a} \frac{1}{a^2}\, dS = \frac{1}{a^2}\,(\text{surface area of } \tilde{S}_a).$$

33. We have

$$
\begin{aligned}
\nabla \times \mathbf{G} &= \nabla \times \int_0^1 t\mathbf{F}(t\mathbf{r}) \times \mathbf{r}\, dt \quad \text{where } \mathbf{r} = (x, y, z), \\
&= \int_0^1 \nabla \times (t\mathbf{F}(t\mathbf{r}) \times \mathbf{r})\, dt \\
&= \int_0^1 t\nabla \times (\mathbf{F}(t\mathbf{r}) \times \mathbf{r})\, dt \quad \text{since } t \text{ behaves as a constant with respect to } \nabla, \\
&= \int_0^1 t\{\mathbf{F}(t\mathbf{r})\nabla \cdot \mathbf{r} - \mathbf{r}\nabla \cdot \mathbf{F}(t\mathbf{r}) + (\mathbf{r} \cdot \nabla)\mathbf{F}(t\mathbf{r}) - (\mathbf{F}(t\mathbf{r}) \cdot \nabla)\mathbf{r}\}\, dt \quad \text{by the first identity,} \\
&= \int_0^1 t\{3\mathbf{F}(t\mathbf{r}) - \mathbf{r}\nabla \cdot \mathbf{F}(t\mathbf{r}) + (\mathbf{r} \cdot \nabla)\mathbf{F}(t\mathbf{r}) - \mathbf{F}(t\mathbf{r})\}\, dt \\
&= \int_0^1 t\{2\mathbf{F}(t\mathbf{r}) - \mathbf{r}\nabla \cdot \mathbf{F}(t\mathbf{r}) + (\mathbf{r} \cdot \nabla)\mathbf{F}(t\mathbf{r})\}\, dt.
\end{aligned}
$$

To compute $\nabla \cdot \mathbf{F}(t\mathbf{r})$, note that $\dfrac{\partial}{\partial x}\mathbf{F}(t\mathbf{r}) = t\dfrac{\partial \mathbf{F}}{\partial \mathbf{X}}$ by the hint. This implies that $\nabla \cdot \mathbf{F}(t\mathbf{r}) = t\nabla_{\mathbf{X,Y,Z}}\mathbf{F}(\mathbf{X}, \mathbf{Y}, \mathbf{Z})$ where $\nabla_{\mathbf{X,Y,Z}}$ signifies that all partials are to be taken with respect to **X**, **Y**, and **Z** where $\mathbf{X} = tx$, $\mathbf{Y} = ty$, and $\mathbf{Z} = tz$. Thus $\nabla \cdot \mathbf{F}(t\mathbf{r}) = 0$ since **F** is assumed to be divergenceless. By the second identity given in the hint,

$$
(\mathbf{r} \cdot \nabla)\mathbf{F}(t\mathbf{r}) = \frac{d}{dt}[t\mathbf{F}(t\mathbf{r})] - \mathbf{F}(t\mathbf{r}).
$$

Hence,

$$
\begin{aligned}
\nabla \times \mathbf{G} &= \int_0^1 t\left\{2\mathbf{F}(t\mathbf{r}) + \frac{d}{dt}[t\mathbf{F}(t\mathbf{r})] - \mathbf{F}(t\mathbf{r})\right\} dt \\
&= \int_0^1 t\left\{\mathbf{F}(t\mathbf{r}) + \frac{d}{dt}[t\mathbf{F}(t\mathbf{r})]\right\} dt \\
&= \int_0^1 \frac{d}{dt}[t^2\mathbf{F}(t\mathbf{r})]\, dt \quad \text{by the last identity in the hint,} \\
&= t^2\mathbf{F}(t\mathbf{r})\Big|_{t=0}^{1} = \mathbf{F}(\mathbf{r}).
\end{aligned}
$$

37. Since $\nabla \times (\nabla\phi) = \mathbf{0}$ for any C^2 function, we have

$$
\nabla \times (\mathbf{G} + \nabla\phi) = \nabla \times \mathbf{G} + \nabla \times (\nabla\phi) = \nabla \times \mathbf{G} + \mathbf{0} = \mathbf{F}.
$$

Thus $\mathbf{G} + \nabla\phi$ is a vector potential for **F**.

41. We have $\mathbf{E} = \nabla f - \dfrac{\partial \mathbf{A}}{\partial t}$ so that Gauss's law becomes $\rho/\epsilon_0 = \nabla \cdot \mathbf{E} = \nabla \cdot \left(\nabla f - \dfrac{\partial \mathbf{A}}{\partial t}\right) = \nabla^2 f - \dfrac{\partial}{\partial t}(\nabla \cdot \mathbf{A})$ or $\nabla^2 f = \rho/\epsilon_0 + \dfrac{\partial}{\partial t}(\nabla \cdot \mathbf{A})$.

Chapter 8

Vector Analysis in Higher Dimensions

8.1 An Introduction to Differential Forms

1. $(dx_1 - 3\,dx_2)(7,3) = dx_1(7,3) - 3\,dx_2(7,3) = 7 - 3(3) = -2.$

7.

$$\begin{aligned}
&(2\,dx_1 \wedge dx_3 \wedge dx_4 + dx_2 \wedge dx_3 \wedge dx_5)(\mathbf{a}, \mathbf{b}, \mathbf{c}) \\
&= 2\det\begin{bmatrix} dx_1(\mathbf{a}) & dx_1(\mathbf{b}) & dx_1(\mathbf{c}) \\ dx_3(\mathbf{a}) & dx_3(\mathbf{b}) & dx_3(\mathbf{c}) \\ dx_4(\mathbf{a}) & dx_4(\mathbf{b}) & dx_4(\mathbf{c}) \end{bmatrix} + \det\begin{bmatrix} dx_2(\mathbf{a}) & dx_2(\mathbf{b}) & dx_2(\mathbf{c}) \\ dx_3(\mathbf{a}) & dx_3(\mathbf{b}) & dx_3(\mathbf{c}) \\ dx_5(\mathbf{a}) & dx_5(\mathbf{b}) & dx_5(\mathbf{c}) \end{bmatrix} \\
&= 2\det\begin{bmatrix} 1 & 0 & 5 \\ -1 & 9 & 0 \\ 4 & 1 & 0 \end{bmatrix} + \det\begin{bmatrix} 0 & 0 & 0 \\ -1 & 9 & 0 \\ 2 & -1 & -2 \end{bmatrix} \\
&= 2(-185) + 0 = -370.
\end{aligned}$$

11.

$$\begin{aligned}
\omega_{(x,y,z)}((2,0,-1),\ (1,7,5)) &= \cos x \begin{vmatrix} 2 & 1 \\ 0 & 7 \end{vmatrix} - \sin z \begin{vmatrix} 0 & 7 \\ -1 & 5 \end{vmatrix} + (y^2+3)\begin{vmatrix} 2 & 1 \\ -1 & 5 \end{vmatrix} \\
&= 14\,\cos x - 7\sin z + 11(y^2+3)
\end{aligned}$$

15. From Definition 1.3 of exterior product,

$$\begin{aligned}
&(y\,dx - x\,dy) \wedge (z\,dx \wedge dy + y\,dx \wedge dz + x\,dy \wedge dz) \\
&\quad = yz\,dx \wedge dx \wedge dy - xz\,dy \wedge dx \wedge dy + y^2\,dx \wedge dx \wedge dz - xy\,dy \wedge dx \wedge dz \\
&\qquad + xy\,dx \wedge dy \wedge dz - x^2\,dy \wedge dy \wedge dz \\
&\quad = 2xy\,dx \wedge dy \wedge dz \quad \text{using (3) and (4).}
\end{aligned}$$

17. Again from Definition 1.3 of exterior product,

$$\begin{aligned}
&(x_1\,dx_1 + 2x_2\,dx_2 + 3x_3\,dx_3) \wedge ((x_1+x_2)\,dx_1 \wedge dx_2 \wedge dx_3 + (x_3 - x_4)\,dx_1 \wedge dx_2 \wedge dx_4) \\
&\quad = x_1(x_1+x_2)\,dx_1 \wedge dx_1 \wedge dx_2 \wedge dx_3 + 2x_2(x_1+x_2)\,dx_2 \wedge dx_1 \wedge dx_2 \wedge dx_3 \\
&\qquad + 3x_2(x_1+x_2)\,dx_3 \wedge dx_1 \wedge dx_2 \wedge dx_3 + x_1(x_3-x_4)\,dx_1 \wedge dx_1 \wedge dx_2 \wedge dx_4 \\
&\quad = 2x_2(x_3-x_4)\,dx_2 \wedge dx_1 \wedge dx_2 \wedge dx_4 + 3x_3(x_3-x_4)\,dx_3 \wedge dx_1 \wedge dx_2 \wedge dx_4.
\end{aligned}$$

Using equation (4), this last expression is equal to

$$0+0+0+0+0+3x_3(x_3-x_4)\,dx_3 \wedge dx_1 \wedge dx_2 \wedge dx_4 = 3x_3(x_3-x_4)\,dx_1 \wedge dx_2 \wedge dx_3 \wedge dx_4,$$

using equation (3).

21. This is easier to show in person, but the point is that if you switch the two identical forms then, on the one hand, nothing has changed and, on the other hand, formula (3) says that you now have the negative of what you started with. So

$$dx_{i_1} \wedge dx_{i_2} \wedge \cdots \wedge dx_{i_j} \wedge \cdots \wedge dx_{i_j} \wedge \cdots \wedge dx_{i_k} = -dx_{i_1} \wedge dx_{i_2} \wedge \cdots \wedge dx_{i_j} \wedge \cdots \wedge dx_{i_j} \wedge \cdots \wedge dx_{i_k}$$

and therefore

$$dx_{i_1} \wedge dx_{i_2} \wedge \cdots \wedge dx_{i_j} \wedge \cdots \wedge dx_{i_j} \wedge \cdots \wedge dx_{i_k} = 0.$$

23. Let $\omega_1 = \sum F_{i_1 \dots i_k}\, dx_{i_1} \wedge \cdots \wedge dx_{i_k}$, $\omega_2 = \sum G_{i_1 \dots i_k}\, dx_{i_1} \wedge \cdots \wedge dx_{i_k}$, and $\eta = \sum H_{j_1 \dots j_l}\, dx_{j_1} \wedge \cdots \wedge dx_{j_l}$. Then

$$\begin{aligned}
(\omega_1 + \omega_2) \wedge \eta &= \left[\sum_{i_1,\dots,i_k} (F_{i_1 \dots i_k} + G_{i_1 \dots i_k})\, dx_{i_1} \wedge \cdots \wedge dx_{i_k} \right] \wedge \sum_{j_1,\dots,j_l} H_{j_1 \dots j_l}\, dx_{j_1} \wedge \cdots \wedge dx_{j_l} \\
&= \sum_{\substack{i_1,\dots,i_k \\ j_1,\dots,j_l}} (F_{i_1 \dots i_k} + G_{i_1 \dots i_k}) H_{j_1 \dots j_l}\, dx_{i_1} \wedge \cdots \wedge dx_{i_k} \wedge dx_{j_1} \wedge \cdots \wedge dx_{j_l} \\
&= \sum_{\substack{i_1,\dots,i_k \\ j_1,\dots,j_l}} F_{i_1 \dots i_k} H_{j_1 \dots j_l}\, dx_{i_1} \wedge \cdots \wedge dx_{i_k} \wedge dx_{j_1} \wedge \cdots \wedge dx_{j_l} \\
&\quad + \sum_{\substack{i_1,\dots,i_k \\ j_1,\dots,j_l}} G_{i_1 \dots i_k} H_{j_1 \dots j_l}\, dx_{i_1} \wedge \cdots \wedge dx_{i_k} \wedge dx_{j_1} \wedge \cdots \wedge dx_{j_l} \\
&= \omega_1 \wedge \eta + \omega_2 \wedge \eta.
\end{aligned}$$

25. Let $\omega = \sum F_{i_1 \dots i_k}\, dx_{i_1} \wedge \cdots \wedge dx_{i_k}$, $\eta = \sum G_{j_1 \dots j_l}\, dx_{j_1} \wedge \cdots \wedge dx_{j_l}$, and $\tau = \sum H_{u_1 \dots u_m}\, dx_{u_1} \wedge \cdots \wedge dx_{u_m}$. Then

$$\begin{aligned}
(\omega \wedge \eta) \wedge \tau &= \left[\sum_{\substack{i_1,\dots,i_k \\ j_1,\dots,j_l}} F_{i_1 \dots i_k} G_{j_1 \dots j_l}\, dx_{i_1} \wedge \cdots \wedge dx_{i_k} \wedge dx_{j_1} \wedge \cdots \wedge dx_{j_l} \right] \\
&\quad \wedge \sum_{u_1,\dots,u_m} H_{u_1 \dots u_m}\, dx_{u_1} \wedge \cdots \wedge dx_{u_m} \\
&= \sum_{\substack{i_1,\dots,i_k \\ j_1,\dots,j_l \\ u_1,\dots,u_m}} F_{i_1 \dots i_k} G_{j_1 \dots j_l} H_{u_1 \dots u_m}\, dx_{i_1} \wedge \cdots \wedge dx_{i_k} \wedge dx_{j_1} \wedge \cdots \wedge dx_{j_l} \wedge dx_{u_1} \wedge \cdots \wedge dx_{u_m}.
\end{aligned}$$

Similarly, calculate $\omega \wedge (\eta \wedge \tau)$ and you will obtain the same result.

8.2 Manifolds and Integrals of k-Forms

3. This is a combination of Example 3 and Exercise 2. Let's begin by describing the location of the point (x_1, y_1). It is anywhere on a circle of radius 3 centered at the origin. So $(x_1, y_1) = (3\cos\theta_1, 3\sin\theta_1)$ where $0 \le \theta_1 < 2\pi$. We can then describe (x_2, y_2) as being this same annular region centered at (x_1, y_1). Together this means that the locus of (x_2, y_2) is $(3\cos\theta_1 + l_2\cos\theta_2, 3\sin\theta_1 + l_2\sin\theta_2)$ where $1 \le l_2 \le 2$ and $0 \le \theta_2 < 2\pi$. Similarly we describe (x_3, y_3) in terms of (x_2, y_2) using variables l_3 and θ_3 such that $1 \le l_3 \le 2$ and $0 \le \theta_2 < 2\pi$. The mapping is

$$\begin{aligned}
\mathbf{X}(\theta_1, l_2, \theta_2, l_3, \theta_3) = (&3\cos\theta_1, 3\sin\theta_1, 3\cos\theta_1 + l_2\cos\theta_2, 3\sin\theta_1 + l_2\sin\theta_2, \\
&3\cos\theta_1 + l_2\cos\theta_2 + l_3\cos\theta_3, 3\sin\theta_1 + l_2\sin\theta_2 + l_3\sin\theta_3).
\end{aligned}$$

As before, the component functions are at least C^1 so the mapping is at least C^1. As for one-one, consider $\mathbf{X}(\theta_1, l_2, \theta_2, l_3, \theta_3) = \mathbf{X}(\hat{\theta}_1, \hat{l}_2, \hat{\theta}_2, \hat{l}_3, \hat{\theta}_3)$. From the first two component functions we see that $\cos\theta_1 = \cos\hat{\theta}_1$ and $\sin\theta_1 = \sin\hat{\theta}_1$ and $0 \le \theta_1, \hat{\theta}_1 < 2\pi$ so $\theta_1 = \hat{\theta}_1$. Now, (x_2, y_2) lies on a circle of radius l_2 and $(\hat{x}_2, \hat{y}_2)$ lies on a circle of radius $\hat{l}_2$ with each circle centered at the same point $(x_1, y_1) = (\hat{x}_1, \hat{y}_1)$. So $l_2 = \hat{l}_2$. Then, as above, $\cos\theta_2 = \cos\hat{\theta}_2$ and $\sin\theta_2 = \sin\hat{\theta}_2$. As $0 \le \theta_2, \hat{\theta}_2 < 2\pi$, we see that $\theta_2 = \hat{\theta}_2$. Now the rest of the argument follows in exactly the same way since (x_3, y_3) is related to (x_2, y_2) in the same way that (x_2, y_2) is related to (x_1, y_1).

We now need to show that the five tangent vectors are linearly independent.

$$\begin{aligned}
\mathbf{T}_{\theta_1} &= (-3\sin\theta_1, 3\cos\theta_1, -3\sin\theta_1, 3\cos\theta_1, -3\sin\theta_1, 3\cos\theta_1)\\
\mathbf{T}_{l_2} &= (0, 0, \cos\theta_2, \sin\theta_2, \cos\theta_2, \sin\theta_2)\\
\mathbf{T}_{\theta_2} &= (0, 0, -l_2\sin\theta_2, l_2\cos\theta_2, -l_2\sin\theta_2, l_2\cos\theta_2)\\
\mathbf{T}_{l_3} &= (0, 0, 0, 0, \cos\theta_3, \sin\theta_3)\\
\mathbf{T}_{\theta_3} &= (0, 0, 0, 0, -l_3\sin\theta_3, l_3\cos\theta_3)
\end{aligned}$$

Look at the equation $c_1\mathbf{T}_{\theta_1} + c_2\mathbf{T}_{l_2} + c_3\mathbf{T}_{\theta_2} + c_4\mathbf{T}_{l_3} + c_5\mathbf{T}_{\theta_3} = \mathbf{0}$. Because of the leading pair of zeros in all but the vector $\mathbf{T}_{\theta_1}$ we conclude that $c_1 = 0$. The remainder of the argument is exactly as in Exercise 2. Because the first four components of $\mathbf{T}_{l_3}$ and $\mathbf{T}_{\theta_3}$ are zero, we can see that $c_2\cos\theta_2 = c_3 l_2\sin\theta_2$ and $c_2\sin\theta_2 = -c_3 l_2\cos\theta_2$. Solve for c_2 in the first equation and substitute into the second equation to get $c_3 l_2\sin^2\theta_2 = -c_3 l_2\cos^2\theta_2$. Because l_2 cannot be zero, $c_3 = 0$. This then implies that $c_2 = 0$. Given that, we can make the same argument to show $c_4 = c_5 = 0$. Therefore the five tangent vectors are linearly independent and we have described the states of the robot arm as a smooth parametrized 5-manifold in $\mathbf{R}^6$.

7. Parametrize the unit circle C by $\mathbf{x}(t) = (\cos t, \sin t), 0 \le t \le 2\pi$. Then

$$\begin{aligned}
\int_C \omega &= \int_0^{2\pi} \omega_{\mathbf{x}(t)}(-\sin t, \cos t)\,dt = \int_0^{2\pi} (\sin t\,dx - \cos t\,dy)(-\sin t, \cos t)\,dt\\
&= \int_0^{2\pi} (-\sin^2 t - \cos^2 t)\,dt = \int_0^{2\pi} -1\,dt = -2\pi.
\end{aligned}$$

11. (a) For the parametrization given, we calculate the tangent vectors as $\mathbf{T}_{u_1} = (\cos u_2, \sin u_2, 0)$, $\mathbf{T}_{u_2} = (-u_1\sin u_2, u_1\cos u_2, 0)$, and $\mathbf{T}_{u_3} = (0, 0, 1)$. Then

$$\Omega_{\mathbf{X}(u)}(\mathbf{T}_{u_1}, \mathbf{T}_{u_2}, \mathbf{T}_{u_3}) = \det\begin{bmatrix} \cos u_2 & -u_1\sin u_2 & 0\\ \sin u_2 & u_1\cos u_2 & 0\\ 0 & 0 & 1 \end{bmatrix} = u_1.$$

As $0 \le u_1 \le \sqrt{5}$, this is positive when $u_1 \ne 0$. Note that when $u_1 = 0$ the parametrization is not one-one and also that $\mathbf{T}_{u_2} = \mathbf{0}$ so $\mathbf{T}_{u_1}, \mathbf{T}_{u_2}$, and $\mathbf{T}_{u_3}$ are not linearly independent. In other words, the parametrization is not smooth when $u_1 = 0$. It is, however, smooth when $u_1 \ne 0$. You can easily see that the mapping is one-one and at least C^1. To see that the tangent vectors are linearly independent, consider the equation $c_1\mathbf{T}_{u_1} + c_2\mathbf{T}_{u_2} + c_3\mathbf{T}_{u_3} = \mathbf{0}$. We see from the third components that $c_3 = 0$. Look at the remaining equations and we see that

$$\begin{cases} (\cos u_2)c_1 - (u_1\sin u_2)c_2 = 0\\ (\sin u_2)c_1 + (u_1\cos u_2)c_2 = 0. \end{cases}$$

Multiply the first equation by $-\sin u_2$ and the second by $\cos u_2$ and add to obtain $u_1 c_2 = 0$. Because we are assuming that $u_1 \ne 0$, this implies that $c_2 = 0$ and therefore $c_1 = 0$. This shows that the tangent vectors are linearly independent and hence the parametrization is smooth when $u_1 \ne 0$. The conclusion is then that the parametrization given is compatible with the orientation when it is smooth.

(b) We can read the boundary pieces right off of the original parametrization: they are paraboloids that intersect at $z = -1$ in a circle in the plane $z = -1$ of radius $\sqrt{5}$ centered at $(0, 0, -1)$. The boundary is

$$\partial M = \{(x, y, z) | z = x^2 + y^2 - 6, z \le -1\} \cup \{(x, y, z) | z = 4 - x^2 - y^2, z \ge -1\}.$$

We can easily adapt the parametrization to each of these pieces. For the bottom, use

$$\mathbf{Y}_1 : [0, \sqrt{5}] \times [0, 2\pi) \to \mathbf{R}^3; \quad \mathbf{Y}_1(s_1, s_2) = (s_1\cos s_2, s_1\sin s_2, s_1^2 - 6).$$

For the top, use

$$\mathbf{Y}_2 : [0, \sqrt{5}] \times [0, 2\pi) \to \mathbf{R}^3; \quad \mathbf{Y}_2(s_1, s_2) = (s_1\cos s_2, s_1\sin s_2, 4 - s_1^2).$$

(c) On the bottom part of ∂M the outward-pointing unit vector

$$\mathbf{V}_1 = \frac{(2x, 2y, -1)}{\sqrt{4x^2 + 4y^2 + 1}}. \text{ In terms of } \mathbf{Y}_1, \text{ this is } \mathbf{V}_1 = \frac{(2s_1\cos s_2, 2s_1\sin s_2, -1)}{\sqrt{4s_1^2 + 1}}.$$

On the top part of ∂M the outward-pointing unit vector

$$\mathbf{V}_2 = \frac{(2x, 2y, 1)}{\sqrt{4x^2 + 4y^2 + 1}}. \text{ In terms of } \mathbf{Y}_2, \text{ this is } \mathbf{V}_2 = \frac{(2s_1 \cos s_2, 2s_1 \sin s_2, 1)}{\sqrt{4s_1^2 + 1}}.$$

15. We have, for the given parametrization, that $\mathbf{T}_{u_1} = (1, 0, 0, 4(2u_1 - u_3)), \mathbf{T}_{u_2} = (0, 1, 0, 0)$, and $\mathbf{T}_{u_3} = (0, 0, 1, 2(u_3 - 2u_1))$. Thus,

$$\begin{aligned}\omega_{\mathbf{X}(u_1,u_2,u_3)}(\mathbf{T}_{u_1}, \mathbf{T}_{u_2}, \mathbf{T}_{u_3}) &= (u_2\, dx_2 \wedge dx_3 \wedge dx_4 + 2u_1u_3\, dx_1 \wedge dx_2 \wedge dx_3)(\mathbf{T}_{u_1}, \mathbf{T}_{u_2}, \mathbf{T}_{u_3}) \\ &= u_2 \begin{vmatrix} 0 & 1 & 0 \\ 0 & 0 & 1 \\ 4(2u_1 - u_3) & 0 & 2(u_3 - 2u_1) \end{vmatrix} + 2u_1u_3 \begin{vmatrix} 1 & 0 & 0 \\ 0 & 1 & 0 \\ 0 & 0 & 1 \end{vmatrix} \\ &= u_2(8u_1 - 4u_3) + 2u_1u_3 = 8u_1u_2 - 4u_2u_3 + 2u_1u_3.\end{aligned}$$

Hence,

$$\begin{aligned}\int_{\mathbf{X}} \omega &= \int_0^1 \int_0^1 \int_0^1 (8u_1u_2 - 4u_2u_3 + 2u_1u_3)\, du_1\, du_2\, du_3 \\ &= \int_0^1 \int_0^1 (4u_2 - 4u_2u_3 + u_3)\, du_2\, du_3 \\ &= \int_0^1 (2 - 2u_3 + u_3)\, du_3 = 2 - \frac{1}{2} = \frac{3}{2}.\end{aligned}$$

8.3 The Generalized Stokes's Theorem

1. Using Definition 3.1,

$$d(e^{xyz}) = \frac{\partial}{\partial x}(e^{xyz})\, dx + \frac{\partial}{\partial y}(e^{xyz})\, dy + \frac{\partial}{\partial z}(e^{xyz})\, dz + e^{xyz}(yz\, dx + xz\, dy + xy\, dz).$$

5. Again, using Definition 3.1,

$$\begin{aligned}d(xz\, dx \wedge dy - y^2 z\, dx \wedge dz) &= (z\, dx + x\, dz) \wedge dx \wedge dy - (2yz\, dy + y^2\, dz) \wedge dx \wedge dz \\ &= x\, dz \wedge dx \wedge dy - 2yz\, dy \wedge dx \wedge dz \quad \text{using (4) from Section 8.1,} \\ &= (x + 2yz)\, dx \wedge dy \wedge dz \quad \text{using (3) from Section 8.1.}\end{aligned}$$

9. For $\omega = F(x, z)\, dy + G(x, y)\, dz$, we have $d\omega = (F_x\, dx + F_z\, dz) \wedge dy + (G_x\, dx + G_y\, dy) \wedge dz$. Expanding, this gives $d\omega = F_x\, dx \wedge dy + G_x\, dx \wedge dz + (G_y - F_z)\, dy \wedge dz$. But we are told that $d\omega = z\, dx \wedge dy + y\, dx \wedge dz$ so

$$\frac{\partial F}{\partial x} = z, \frac{\partial G}{\partial x} = y, \text{ and } \frac{\partial G}{\partial y} - \frac{\partial F}{\partial z} = 0.$$

The first equation implies that $F(x, z) = xz + f(z)$ for some differentiable function f of z alone. Similarly, the second equation implies that $G(x, y) = xy + g(y)$ for some differentiable function g of y alone. Using these results together with the third equation we see that $x + g'(y) = x + f'(z)$ or $g'(y) = f'(z)$. This can only be true if their common value is a constant C. So if $g'(y) = f'(z) = C$, then $f(z) = Cz + D_1$ and $g(y) = Cy + D_2$ for arbitrary constants C, D_1, and D_2. We conclude that $F(x, z) = xz + Cz + D_1$ and $G(x, y) = xy + Cy + D_2$.

11. One integral is easy. Since $\omega = xy\,dz \wedge dw$ and $\partial M = \{(x,y,z,w)|x = 0, 8 - 2y^2 - 2z^2 - 2w^2 = 0\}$, we see that $x = 0$ along ∂M so $\int_{\partial M} \omega = \int_{\partial M} 0 = 0$.

Now $d\omega = d(xy) \wedge dz \wedge dw = x\,dy \wedge dz \wedge dw + y\,dx \wedge dz \wedge dw$. We can orient M any way we wish, so we won't worry about this—we'll choose the orientation to be compatible with the parametrization.

$$\mathbf{X} : D \to \mathbf{R}^4, \quad \mathbf{X}(u_1, u_2, u_3) = (8 - 2u_1^2 - 2u_2^2 - 2u_3^2, u_1, u_2, u_3)$$

where $D = \{(u_1, u_2, u_3)|u_1^2 + u_2^2 + u_3^2 \leq 4\}$ (i.e., the solid ball of radius 2). Then

$$\begin{aligned}
\int_M d\omega &= \int_{\mathbf{X}} d\omega = \iiint_B d\omega_{\mathbf{X}(u)}(\mathbf{T}_{u_1}, \mathbf{T}_{u_2}, \mathbf{T}_{u_3})\,du_1\,du_2\,du_3 \\
&= \iiint_B \left\{ (8 - 2u_1^2 - 2u_2^2 - 2u_3^2) \begin{vmatrix} 1 & 0 & 0 \\ 0 & 1 & 0 \\ 0 & 0 & 1 \end{vmatrix} + u_1 \begin{vmatrix} -4u_1 & -4u_2 & -4u_3 \\ 0 & 1 & 0 \\ 0 & 0 & 1 \end{vmatrix} \right\} du_1\,du_2\,du_3 \\
&= \iiint_B (8 - 2(u_1^2 + u_2^2 + u_3^2) - 4u_1^2)\,du_1\,du_2\,du_3.
\end{aligned}$$

At this point it is helpful to switch to spherical coordinates. The previous quantity is then

$$\begin{aligned}
&= \int_0^{2\pi} \int_0^{\pi} \int_0^2 (8 - 2\rho^2 - 4\rho^2 \sin^2\varphi \cos^2\theta)\rho^2 \sin\varphi\,d\rho\,d\varphi\,d\theta \\
&= 8 \cdot (\text{volume of } B) - 2\int_0^{2\pi} \int_0^{\pi} \int_0^2 \rho^4(\sin\varphi + 2\sin^3\varphi\cos^2\theta)\,d\rho\,d\varphi\,d\theta \\
&= 8 \cdot \left(\frac{4}{3}\pi 2^3\right) - 2\int_0^{2\pi} \int_0^{\pi} \frac{32}{5}(\sin\varphi + 2\sin\varphi(1 - \cos^2\varphi)\cos^2\theta)\,d\varphi\,d\theta \\
&= \frac{256\pi}{3} - \frac{64}{5}\int_0^{2\pi} \{(-\cos\varphi)\Big|_0^{\pi} + 2\cos^2\theta(-\cos\varphi + (\cos^3\varphi)/3)\Big|_{\varphi=0}^{\pi}\}\,d\theta \\
&= \frac{256\pi}{3} - \frac{64}{5}\int_0^{2\pi} \left\{2 + 2\cos^2\theta \cdot \left(2 - \frac{2}{3}\right)\right\} d\theta = \frac{256\pi}{3} - \frac{128}{5}\int_0^{2\pi} \left(1 + \frac{4}{3}\cos^2\theta\right) d\theta \\
&= \frac{256\pi}{3} - \frac{128}{5}\int_0^{2\pi} \left(\frac{5}{3} + \frac{2}{3}\cos 2\theta\right) d\theta \quad \text{(using the half angle formula)} \\
&= \frac{256\pi}{3} - \frac{128}{5}\left(\frac{5}{3}(2\pi) + \frac{1}{3}\sin 2\theta\Big|_0^{2\pi}\right) = \frac{256\pi}{3} - \frac{256\pi}{3} = 0.
\end{aligned}$$

True/False Exercises for Chapter 8

1. True.
3. True.
5. True.
7. True.
9. True.
11. False. ($\mathbf{X}(1,1,-1) = \mathbf{X}(1,1,1)$, so $\mathbf{X}$ is not one-one on D.)
13. False. (The agreement is only up to sign.)
15. False. (This is only true if n is even.)
17. True.
19. True. ($d\omega$ would be an $(n+1)$-form, and there are no nonzero ones on $\mathbf{R}^n$.)

Miscellaneous Exercises for Chapter 8

1. (a) First, by definition of the exterior product and derivative

$$\begin{aligned} d(f \wedge g) &= \sum_{i=1}^{n} \frac{\partial}{\partial x_i}(fg)\, dx_i = \sum_{i=1}^{n} \left(\frac{\partial f}{\partial x_i} g + f \frac{\partial g}{\partial x_i} \right) dx_i \quad \text{by the product rule,} \\ &= g \sum_{i=1}^{n} \frac{\partial f}{\partial x_i}\, dx_i + f \sum_{i=1}^{n} \frac{\partial g}{\partial x_i}\, dx_i \\ &= g \wedge df + f \wedge dg \\ &= df \wedge g + (-1)^0 f \wedge dg. \end{aligned}$$

(b) If $k = 0$, then write $\omega = f$ so that

$$\begin{aligned} d(\omega \wedge \eta) &= d(f \wedge \eta) = d\left(\sum f G_{j_1 \dots j_l}\, dx_{j_1} \wedge \cdots \wedge dx_{j_l} \right) = \sum d(f G_{j_1 \dots j_l}) \wedge dx_{j_1} \wedge \cdots \wedge dx_{j_l} \\ &= \sum (df \wedge G_{j_1 \dots j_l} + f \wedge dG_{j_1 \dots j_l}) \wedge dx_{j_1} \wedge \cdots \wedge dx_{j_l} \quad \text{from (a),} \\ &= df \wedge \sum G_{j_1 \dots j_l}\, dx_{j_1} \wedge \cdots \wedge dx_{j_l} + f \wedge \sum dG_{j_1 \dots j_l}\, dx_{j_1} \wedge \cdots \wedge dx_{j_l} \\ &= df \wedge \eta + (-1)^0 f \wedge d\eta. \end{aligned}$$

(c) If $l = 0$, then write $\eta = g$ so that

$$\begin{aligned} d(\omega \wedge \eta) &= d(\omega \wedge g) = d\left(\sum g F_{i_1 \dots i_k}\, dx_{i_1} \wedge \cdots \wedge dx_{i_k} \right) \\ &= \sum d(g F_{i_1 \dots i_k}) \wedge dx_{i_1} \wedge \cdots \wedge dx_{i_k} \\ &= \sum (dg \wedge F_{i_1 \dots i_k} + g \wedge dF_{i_1 \dots i_k}) \wedge dx_{i_1} \wedge \cdots \wedge dx_{i_k} \\ &= dg \wedge \omega + g \wedge d\omega \\ &= (-1)^k \omega \wedge dg + d\omega \wedge g \end{aligned}$$

by part 2 of Proposition 1.4 (recall dg is a 1-form).

(d) In general,

$$d(\omega \wedge \eta) = d\left(\sum_{1 \le i_1 < \cdots < i_k \le n} F_{i_1 \ldots i_k}\, dx_{i_1} \wedge \cdots \wedge dx_{i_k} \wedge \sum_{1 \le j_1 < \cdots < j_l \le n} G_{j_1 \ldots j_l}\, dx_{j_1} \wedge \cdots \wedge dx_{j_l} \right)$$

$$= d\left(\sum_{\substack{1 \le i_1 < \cdots < i_k \le n \\ 1 \le j_1 < \cdots < j_l \le n}} F_{i_1 \ldots i_k} G_{j_1 \ldots j_l}\, dx_{i_1} \wedge \cdots \wedge dx_{i_k} \wedge dx_{j_1} \wedge \cdots \wedge dx_{j_l} \right)$$

$$= \sum d(F_{i_1 \ldots i_k} G_{j_1 \ldots j_l}) \wedge dx_{i_1} \wedge \cdots \wedge dx_{i_k} \wedge dx_{j_1} \wedge \cdots \wedge dx_{j_l} \quad \text{so by part (a),}$$

$$= \sum (dF_{i_1 \ldots i_k} \wedge G_{j_1 \ldots j_l} + F_{i_1 \ldots i_k} \wedge dG_{j_1 \ldots j_l}) \wedge dx_{i_1} \wedge \cdots \wedge dx_{i_k} \wedge dx_{j_1} \wedge \cdots \wedge dx_{j_l}$$

$$= \sum dF_{i_1 \ldots i_k} \wedge G_{j_1 \ldots j_l} \wedge dx_{i_1} \wedge \cdots \wedge dx_{i_k} \wedge dx_{j_1} \wedge \cdots \wedge dx_{j_l}$$

$$+ \sum F_{i_1 \ldots i_k} \wedge dG_{j_1 \ldots j_l} \wedge dx_{i_1} \wedge \cdots \wedge dx_{i_k} \wedge dx_{j_1} \wedge \cdots \wedge dx_{j_l}$$

$$= \sum dF_{i_1 \ldots i_k} \wedge dx_{i_1} \wedge \cdots \wedge dx_{i_k} \wedge G_{j_1 \ldots j_l} \wedge dx_{j_1} \wedge \cdots \wedge dx_{j_l}$$

$$+ \sum F_{i_1 \ldots i_k} \wedge (-1)^k\, dx_{i_1} \wedge \cdots \wedge dx_{i_k} \wedge dG_{j_1 \ldots j_l} \wedge dx_{j_1} \wedge \cdots \wedge dx_{j_l}$$

since $G_{j_1 \ldots j_l}$ is a 0-form and $dG_{j_1 \ldots j_l}$ is a 1-form,
$= d\omega \wedge \eta + (-1)^k \omega \wedge d\eta.$

5. If S^4 is the unit 4-sphere in $\mathbf{R}^5$, then let B denote the 5-dimensional unit ball

$$B = \{x_1, x_2, x_3, x_4, x_5) | x_1^2 + x_2^2 + x_3^2 + x_4^2 + x_5^2 \le 1\}.$$

Note that $\partial B = S^4$. Then using the generalized Stokes's theorem, we have

$$\int_{S^4} \omega = \int_B d\omega.$$

For $\omega = x_3\, dx_1 \wedge dx_2 \wedge dx_4 \wedge dx_5 + x_4\, dx_1 \wedge dx_2 \wedge dx_3 \wedge dx_5$ we have $d\omega = dx_3 \wedge dx_1 \wedge dx_2 \wedge dx_4 \wedge dx_5 + dx_4 \wedge dx_1 \wedge dx_2 \wedge dx_3 \wedge dx_5 = dx_1 \wedge \cdots \wedge dx_5 - dx_1 \wedge \cdots \wedge dx_5 = 0$. Hence $\int_{S^4} \omega = \int_B 0 = 0$.

9. Because $\partial M = \emptyset$, the note following Theorem 3.2 advises us to take $\int_{\partial M} \omega \wedge \eta$ to be 0 in the equation $\int_{\partial M} \omega \wedge \eta = \int_M d(\omega \wedge \eta)$. Now substitute the results of Exercise 1 to get

$$0 = \int_M d(\omega \wedge \eta) = \int_M d\omega \wedge \eta + (-1)^k \omega \wedge d\eta = \int_M d\omega \wedge \eta + (-1)^k \int_M \omega \wedge d\eta.$$

Pull this last piece to the other side to obtain the result

$$(-1)^{k+1} \int_M \omega \wedge d\eta = \int_M d\omega \wedge \eta.$$